# Current Research in Animal Palaeopathology

Proceedings of the Second ICAZ Animal
Palaeopathology Working Group Conference

Edited by

## Zora Miklíková
## Richard Thomas

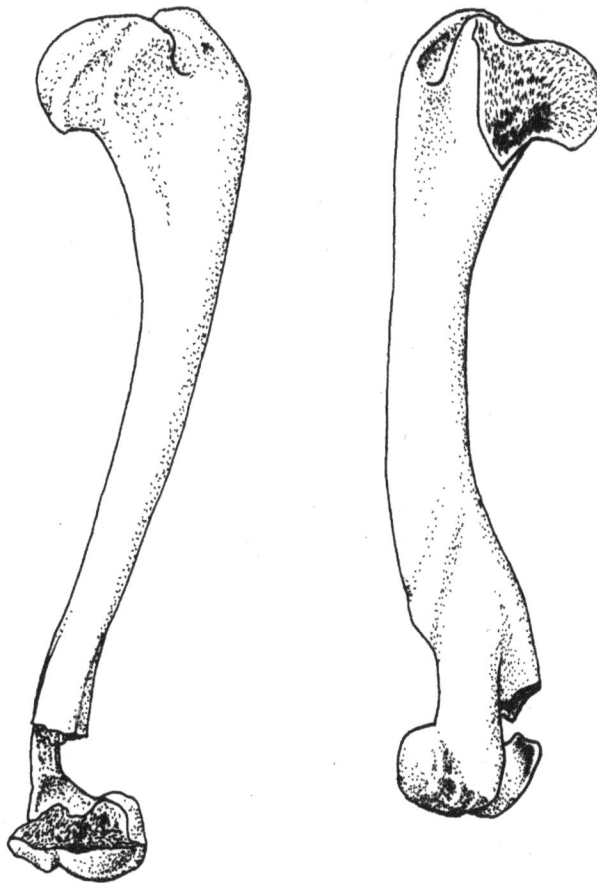

BAR International Series 1844
2008

Published in 2016 by
BAR Publishing, Oxford

BAR International Series 1844

*Current Research in Animal Palaeopathology*

ISBN  978 1 4073 0331 4

BAR Publishing is the trading name of British Archaeological Reports (Oxford) Ltd.
British Archaeological Reports was first incorporated in 1974 to publish the BAR
Series, International and British. In 1992 Hadrian Books Ltd became part of the BAR
group. This volume was originally published by Archaeopress in conjunction with
British Archaeological Reports (Oxford) Ltd / Hadrian Books Ltd, the Series principal
publisher, in 2008. This present volume is published by BAR Publishing, 2016.

Printed in England

# BAR
PUBLISHING

BAR titles are available from:

BAR Publishing
122 Banbury Rd, Oxford, OX2 7BP, UK
EMAIL    info@barpublishing.com
PHONE    +44 (0)1865 310431
FAX      +44 (0)1865 316916
www.barpublishing.com

# Contents

# List of figures

# List of tables

# 1. Introduction: current research in animal palaeopathology

Richard Thomas and Zora Miklíková

Animal palaeopathology can be broadly defined as the study of animal health, disease and injury as primarily revealed through the analysis of animal bones retrieved from archaeological sites. More specifically, it is concerned with those skeletal alterations that result, either directly or indirectly, from disease processes, traumatic insults, nutritional dysfunction, metabolic imbalance, developmental disruptions, and lifestyle. It is not a discipline that belongs solely within the realm of archaeology, however, but instead draws together knowledge across the humanities and sciences, integrating archaeology, history, veterinary medicine, microbiology, and radiology among other disciplines.

The International Council for Archaeozoology (ICAZ) Animal Palaeopathology Working Group (APWG; originally named the Veterinary Pathology Working Group) was set up in 1999 to provide an inter-disciplinary forum for the discussion of theoretical and methodological issues relating to the study of animal health, disease and injury in the past (Thomas and Hammon 1999). At the outset, three specific areas of concern within animal palaeopathological research were identified:

1. a lack of integration of palaeopathological data with other forms of (zoo-)archaeological evidence;
2. inconsistent and/or inadequate recording practices, resulting in the treatment of palaeopathologies as 'interesting specimens', with little concern for geographic and diachronic variation in prevalence;
3. limited understanding of the underlying biological processes and consequences of many different types of pathology.

The papers in this volume arose from the second conference of the APWG, held at the Department of Animal Physiology, Slovak University of Agriculture in Nitra, Slovakia in September 2004. The aim of this conference was to bring together researchers from diverse fields to improve the methodological foundation of the discipline and further demonstrate the value of moving away from the 'interesting specimens' approach, building upon the proceedings of the first APWG conference (Davies *et al.* 2005) in the process.

This volume is organised into two sections reflecting the two distinct themes of the conference. The first nine chapters present site-based or regional analyses of palaeopathological data, and are structured chrono-logically. Three papers deal with prehistoric material (Bartosiewicz, Fabiš *et al.* and Bendrey), four with Roman sites (Vann, Groot, Lyublyanovics and M. Daróczi-Szabó) and two with medieval assemblages (Miklíková and Csippán and L. Daróczi-Szabó). The three papers that constitute the second half of the volume, deal with specific forms of palaeopathological evidence –

trauma, osteoporosis and cranial perforations – and make use of illustrative examples from different periods of time (Gál, Martiniaková *et al.* and Manaseryan).

Each chapter within this volume adds something new to the growing body of animal palaeopathological literature. It is neither appropriate nor necessary to reiterate all of these in detail here, but examples include the identification of conditions that have either not been previously seen (*e.g.* Fabiš *et al.* and Manaseryan) or only rarely observed (*e.g.* tuberculosis: Bendrey; and Daróczi-Szabó) in archaeological assemblages of animal bone.

There are also a number of cross-cutting themes that are worthy of emphasis. Firstly, these papers demonstrate the diversity of archaeological questions upon which animal palaeopathological evidence can inform. Key issues addressed include human attitudes to animals, in life and death, and the impact of different management practices on both animal and human health. The value of palaeopathological evidence in challenging long-held assumptions about human-animal relationships, such as the health impact of animal translocation (Bartosiewicz) or the use of ex-traction cattle for meat supply (Vann), is also clear. Issues of zoological concern can also be revealed through palaeopathological analyses. In this volume, this is exemplified by the evidence for intra-species conflict presented by Gál and the identification of anomalies of skeletal development (*e.g.* Fabiš *et al.* and Manaseryan).

Methodologically, these papers demonstrate the importance of prevalance calculation. Consistent recording of the numbers of pathological and non-pathological specimens permits not only intra- and inter-site analyses (*e.g.* Groot and Vann), but also inter-regional studies (*e.g.* Bartosiewicz). The typically low frequency of anomalies per site and the direct relationship that exists between sample size and pathological frequency (*e.g.* Daróczi-Szabó and Vann), makes the consistent recording of pathological and non-pathological evidence in this way imperative if palaeopathological data are to be used meaningfully. A good example in this volume is provided by the presentation of fracture prevalence for dogs at four Roman sites (Tab 1.1). Continued collation and publication of such data should enable attitudes to these animals to be tracked across space and time.

| Site | Fracture prevalence |
|---|---|
| Alchester | 3.4% |
| Balatonlelle-Kenderføld | 7.4% |
| Colchester | 0% |
| Tiel-Passewaiij | 0.93% |

**Tab 1.1**: frequency of fractured dog bones at four Roman sites.

The importance of making full use of the armoury of techniques and specialists to fully understand particular pathologies is also clear (*e.g.* Manaseryan). While macro-morphological analysis will remain the primary means by which pathological bones are encountered and studied, the continued development and refinemenet of histological (*e.g.* Martiniaková *et al.*), radiographic (*e.g.* O'Connor and O'Connor 2005) and biomolecur analytical techniques (*e.g.* Bendrey *et al.* 2007; Gerneay *et al.* 2006) is significantly improving our understanding of the aetiology and diagnosis of a range of conditions.

In 2000 O'Connor described animal palaeopathology as an "inchoate discipline pursued by a relatively small number of analysts (O'Connor 2000, 98). The studies presented in this volume, combined with the recent research that has been published elsewhere (*e.g.* Bartosiewicz 2006; Becker 2006; Byerly 2007; Izeta and Cortés 2006), indicates that this description is no longer accurate. There are still many areas of the discipline that require methodological advancement (*e.g.* Thomas and Mainland 2005, 4-5); however, animal palaeo-pathological research is burgeoning and is now begininng to make major contributions to our understanding of past human-animal relationships.

## Acknowledgements

In the preparation of this volume there are a number of individuals to whom we are indebted. Firstly we would like to express our thanks to Marián Fabiš, Marcela Kramárová and the rest of the organising committee of the conference for arranging such a stimulating programme and to the Department of Animal Physiology, Slovak University of Agriculture in Nitra, for agreeing to host the meeting. We would also like to thank the contributors for their patience during the production of this volume and to our reviewers for their critical comments.

## Bibliography

Bartosiewicz, L. 2006. Mettre le chariot devant le boeuf. Anomalies ostéologiques liées à l'utilisation des boeuf pour la traction, pp. 259-267, in P. Pétrequin P., Arbogast, R.-M., Péterquin, A.-M., Van Willigen, S. and Bailly, M. (eds), *Premiers Chariots, Premiers Araires. La Diffusion de la Traction Animale en Europe Pendant les IVe et IIIe Millénaires Avant Notre Ère*. CRA Monographies 29. Paris: CNRS Editions.

Becker, I. 2006. Zur nutzungsspezifischen und paläopathologischen Beurteilung von Pferde-knochen aus archäologischen Grabungen – die pferdeskelette von Rullstorf bei Lüneburg. *Beiträge zur Archäozoologie und Prähistorischen Anthropologie* 2, 70-76.

Bendrey, R., Taylor, G. M., Bouwman, A. S. and Cassidy, J. P. 2007. Suspected bacterial disease in two archaeological horse skeletons from southern England: palaeopathological and biomolecular studies. *Journal of Archaeological Science* 35, 1-10.

Byerly, R. M. 2007. Palaeopathology in late Pleistocene and early Holocene Central Plains bison: dental enamel hypoplasia, fluoride toxicosis and the archaeological record. *Journal of Archaeological Science* 34, 1847-1858.

Davies, J., Fabiš, M., Mainland, I., Richards, M. and Thomas, R. 2005. *Diet and Health in Past Animal Populations: Current Research and Future Directions.* Oxford: Oxbow Books.

Gerneay, A. M., Minnikin, D. E., Copley, M. S., Power, J. S., Ahmed, A. M. A., Dixon, R. A., Roberts, C. A., Robertson, D. J., Noland, J. and Chamberlain, A. 2006. Detecting ancient tuberculosis. *Internet Archaeology* 5.

Izeta, A. D. and Cortés, L. I. 2006. South American camelid palaeopathologies: examples from Loma Alta (Catamarca, Argentina). *International Journal of Osteoarchaeology* 16 (3), 269-275.

O'Connor, T. P. 2000. *The Archaeology of Animal Bones*. Stroud: Sutton.

O'Connor, T. P. and O'Connor, S. 2005. Digitising and image-processing radiographs to enhance interpretations in avian palaeopathology, pp. 69-82, in Grupe, G. and Peters, J. (eds), *Feathers, Grit and Symbolism. Birds and Humans in the Ancient Old and New Worlds*. Documenta Archaeobiologae 3. Rahden: Marie Leidorf Verlag.

Thomas, R. and Hammon, A. 1999. Veterinary Palaeo-pathology Working Group. *Organ: the Newsletter of the Osteoarchaeology Research Group* 20, 2-3.

Thomas, R. and Mainland, I. 2005. Introduction: animal diet and health – current perspectives and future directions, pp. 1-7, in Davies, J., Fabiš, M., Mainland, I., Richards, M. and Thomas, R. (eds), *Diet and Health in Past Animal Populations: Current Research and Future Directions*. Oxford: Oxbow.

## Authors' affiliation

Richard Thomas
School of Archaeology and Ancient History
University of Leicester
University Road
Leicester LE1 7RH
United Kingdom

Zora Miklíková
Institute of Archaeology
Slovak Academy of Sciences
Akademická 2
949 21 Nitra
Slovakia

# 2. Environmental stress in early domestic sheep

László Bartosiewicz

## Abstract

*The first domesticates in the Carpathian Basin are known from the early Neolithic Körös culture. Proceeding north-west from the Balkans, sheep and goat first encountered a marshy habitat here. It is hypothesised that pathological deformations on the bones of animals recovered from archaeological sites in this region resulted from environmental stress in the humid habitat of the Great Hungarian Plain. This theory is tested using a broad range of published archaeozoological data to compare the prevalence of two types of pathology – periodontal disease and arthritis – amongst sheep in Central Europe and in their native Near East. Statistical analyses of these data reveal that while oral pathology occurred more commonly at European sites, arthritic deformations developed with greater probability in more arid areas in the Near East. The possible causes of these patterns are discussed.*

## Introduction

The first domestic animals in the Carpathian Basin occurred at sites of the Early Neolithic Körös culture, recently studied in great detail (Whittle *et al.* 2002, 2005). In all likelihood, sheep (*Ovis aries* L., 1758) and goat (*Capra hircus* L., 1758) reached this area by diffusion from the south-east, since their wild ancestors were not available for local domestication in Central Europe. Proceeding north-west from the Balkans out of south-west Asia, these animals first encountered a marshy habitat in the Great Hungarian Plain, an environment which must have posed serious problems for their keeping. This holds equally true for both species, although sometimes their bone fragments cannot be reliably distinguished and they are thus identified by the generic term sheep/goat or caprine (subfamily: Caprinae Gray, 1821). The majority of known remains in Europe, however, usually originate from sheep (the ratio is usually five to one in Hungary; Bartosiewicz 1999), while in the Near East, an increasing trend in goat keeping has been observed to parallel aridity (Bartosiewicz 1984; Bökönyi and Bartosiewicz 2000).

The relatively small size of sheep at the Körös culture settlements of Endrőd 119 (Bökönyi 1992) and Ecsegfalva 23 (Bartosiewicz 2005) in Hungary seems to have been comparable to modern, unimproved Shetland ewes (Davis 1996). When plotted around the mean bone measurements of a reference sample formed by the latter, the resulting bimodal distribution of standard scores for Early Neolithic sheep suggests meagre size and clearly expressed secondary sexual dimorphism (Fig. 2.1). Naturally, this rather striking similarity in size between Neolithic ewes in Hungary and a modern but unimproved breed in Britain should not be seen as any sign of continuity. However, it inspired the hypothesis that the early domestic sheep of south-west Asian origin were exposed to stress in their new environments in Europe which, in addition to their small size, must have resulted in a host of pathological conditions. In this paper, two broadly defined conditions are discussed in detail: periodontal disease and *lato sensu* arthritis.

Symptoms of periodontal disease were first evaluated against the backdrop of published data. The multi-causal aetiology of this condition in herbivores (Haimovici and Haimovici 1971) probably includes overgrazing. The removal of edible grass from limited pastures creates empty areas of ground often filled by rough and thorny plants. These plants gradually take over, increasing the incidence of oral injury and concomitant infections.

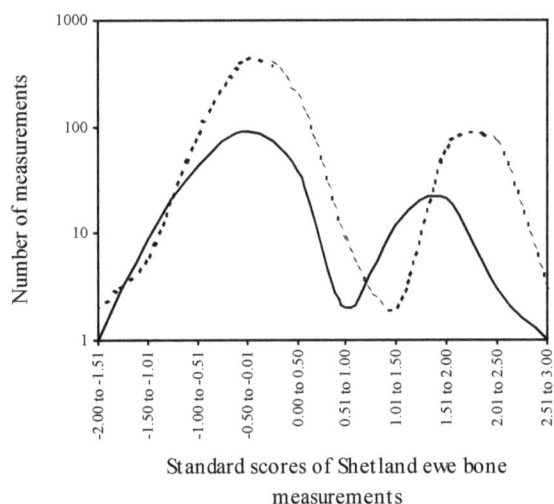

Figure 2.1: bimodal distribution of Early Neolithic sheep bone measurements from Hungary compared with the standard scores of modern Shetland ewes (Davis 1996). Continuous line: Ecsegfalva 23; dashed line: Endrőd 119.

Arthritis was first described in cave bear as *Höhlengicht* ('cave gout'; Virchow 1895). This implies cool cave climates, a possible factor in this condition. Environment, however, also includes nutrition (von den Driesch 1975: 420) and overworking, as well as the lack of exercise (Holmberg and Reiland 1984), all of which are associated with arthritis. Arthritis is also a consequence of age-related decrease in bone vascularisation. Although studies in sheep are yet to be carried out, in modern draught oxen the severity of arthritis has been shown to increase with age: the arthritic fusion between tarsal bones did not occur below eight years and 475 kg in a study conducted by Bartosiewicz *et al.* (1997, Table 39). A number of unknown inflammations, resulting in the formation of morphologically similar lesions, may be caused by metabolic disorders and infections, whose archaeological identification is impossible. Finally, inheritance must be considered. Joint deformations observed in modern fallow deer in Britain (Chaplin 1971, 118; Fig. 17), for example, point to lax selection pressure adding a congenital component to this range of causes.

It would be convenient to attribute both major types of lesions to the environmental stress to which freshly imported caprines were exposed in new habitats in Central Europe. In the case of the Early Neolithic Körös culture, insistence on keeping mostly sheep for some 600 years despite adverse environmental circumstances may reflect a strong cultural tradition which only slowly turned into what would today be considered rational adaptation to a special environment (Bartosiewicz and Choyke 1985). It was therefore worth testing the hypothesis that caprines in Central Europe showed different patterns of morbidity than their kin in the Near East. A targeted review of animal palaeopathology in the latter region has shown that differences in the manifestation of animal disease are attributable to differences in the fauna exploited and environmental conditions as well as assemblage size (Bartosiewicz 2002, 334).

## Materials and Methods

The small concentration of pathologically deformed sheep bones from Early Neolithic sites in Hungary would have made the statistical analysis of this problem impossible. In the first comprehensive study of marginal periodontal lesions in archaeozoology, Haimovici and Haimovici (1971, 261-266, Figs. 1-11) studied 4200 fragmentary tooth rows (70% of which were mandibles) dating from the Late Neolithic to the La Tène Period (third millennium BC to the first century AD) from numerous sites in Romania. Amongst this vast material they found only six affected tooth rows from sheep, two from red deer and only single incidences from cow, pig and horse. Thus, the recovery of two pathological specimens among the 210 caprine mandible fragments from Ecsegfalva 23 may be considered fortuitous, possibly reflecting finer techniques of excavation. Given the rare occurrence of pathological specimens, published literature was consulted for the purposes of this study. In an effort to create samples sufficiently large for statistical analysis sites, two broad geographic units, Central Europe and the Near East, were chosen for study. Materials from various archaeological periods were pooled: breaking down the data set by chronological units, would have resulted in sample sizes far too small for quantitative analysis. Consequently, diachronic aspects of the problem could not be studied. Faunal assemblages from Central Europe were predominantly represented by sites in

Germany, while tell settlements from Turkey and Iran mainly contributed to the Near Eastern dataset. Owing to their complex stratigraphy, these latter sites usually provided several distinct chronological assemblages. A Chi-square test was carried out to verify hypothetical differences between the morbidity of sheep in these two regions.

The majority of cases used in this paper were adopted from the dissertation series of the *Institut für Paläoanatomie, Domestikationsforschung und Geschichte der Tiermedizin der Universität München*. It was hoped that these specimens, identified by veterinary students supervised by the late Prof. Joachim Boessneck and his staff, would provide a relatively homogeneous and reliable source for this investigation. Only cases where the pathological condition could be reasonably well verified were included. Fortunately, due to the rare occurrence of pathological specimens, the pictorial documentation of such bones tends to be rather consistent and detailed. The basic parameters of the two regional samples are summarised in Tab. 2.1.

| | Central Europe | Near East |
|---|---|---|
| Total number of sites | 28 | 8 |
| Known NISP (pooled) | 283,473 (n=13) | 327,184 (n=20) |
| Mean NISP | 21,806 | 16,359 |
| Mean % of caprine NISP | 26.3 | 37.9 |
| Periodontal disease, n | 36 | 14 |
| Arthritic lesions, n | 17 | 25 |

Table 2.1: sites whose pathological caprine bones were used in the analysis (for details see the Appendix).

## Results

Firstly, the anatomical and pathological features of the two major conditions selected for study will be briefly reviewed in the pooled material, *i.e.* with no distinction between Central European and Near Eastern specimens. Hypothetical differences between caprine morbidity in the two areas will then be tested.

## Periodontal disease

Most of the reliable cases of periodontal disease observed in the literature were recorded in the lower tooth row. A typical case on a Neolithic sheep mandible is shown in Fig. 2.2. The anatomical distribution of periodontal symptoms in 29 caprine mandibles is summarised in Tab. 2.2. No identifiable goat mandibles showed this condition in the dataset available for study.

Aside from the region of the second premolar, directly exposed to random external effects in the oral cavity, most symptoms are concentrated between the lower third premolar and the second molar, with a special predisposition of the first lower molar. Hundreds of sheep mandibles from Wicken Bonhunt, a Saxon site in England (Levitan 1977, quoted by Baker and Brothwell 1980, 152-153, Figs. 10-11) showed a similar pattern in

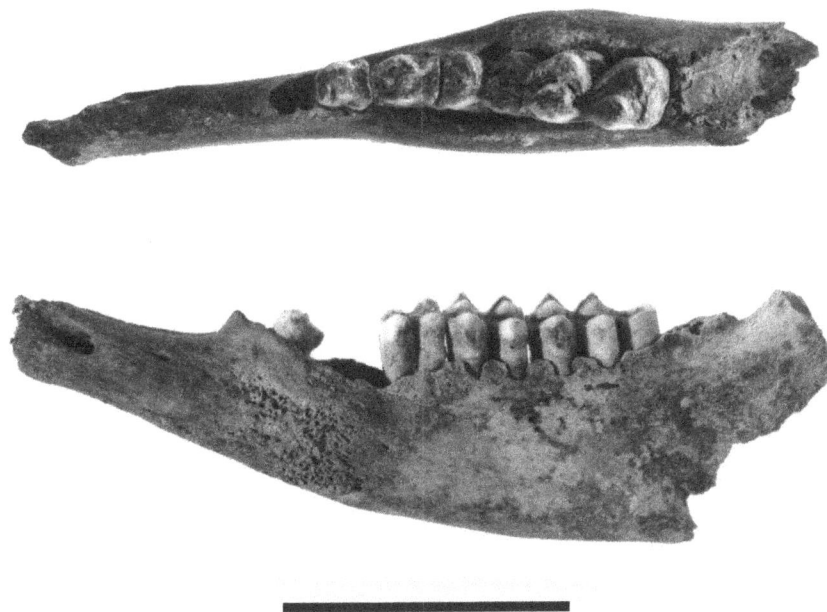

**Figure 2.2**: typical cases of gingivitis in Neolithic sheep mandibles from Karanovo, Bulgaria. Occlusal and buccal views. Scale bar: 50 mm.

| Jaw No. | $P_2$ | $P_3$ | $P_4$ | $M_1$ | $M_2$ | $M_3$ | Total |
|---|---|---|---|---|---|---|---|
| 1 | ▓ | | | | | | 1 |
| 2 | ▓ | | | | | | 1 |
| 3 | ▓ | | | | | | 1 |
| 4 | ▓ | | | | | | 1 |
| 5 | ▓ | | | | | | 1 |
| 6 | ▓ | ▓ | | | | | 2 |
| 7 | ▓ | ▓ | ▓ | ▓ | | | 4 |
| 8 | ▓ | ▓ | ▓ | ▓ | ▓ | | 5 |
| 9 | | | ▓ | | ▓ | | 2 |
| 10 | | ▓ | ▓ | ▓ | ▓ | ▓ | 5 |
| 11 | | | ▓ | ▓ | ▓ | | 3 |
| 12 | | | ▓ | ▓ | | | 2 |
| 13 | | | ▓ | ▓ | | | 2 |
| 14 | | | ▓ | ▓ | | | 2 |
| 15 | | | | ▓ | | | 1 |
| 16 | | | | ▓ | | | 1 |
| 17 | | | | ▓ | | | 1 |
| 18 | | | | ▓ | | | 1 |
| 19 | | | | ▓ | | | 1 |
| 20 | | | | ▓ | | | 1 |
| 21 | | | | ▓ | ▓ | | 2 |
| 22 | | | | ▓ | ▓ | | 2 |
| 23 | | | | ▓ | ▓ | | 2 |
| 24 | | | | ▓ | ▓ | | 2 |
| 25 | | ▓ | | | ▓ | | 2 |
| 26 | | | | | ▓ | | 1 |
| 27 | | | | | | ▓ | 1 |
| 28 | | | | | | ▓ | 1 |
| 29 | | ▓ | | | | | 1 |
| **Total** | **9** | **7** | **9** | **19** | **8** | **3** | **55** |

**Table 2.2**: areas affected by periodontal disease in 29 sheep and sheep/goat mandibles.

the distribution of these lesions. As periodontal disease, caused by alveolar infections, spreads to adjacent tissues it may cause osteomyelitis (Nieberle and Cohrs 1970, 403) and the formation of fistulae (Tamás 1987, 118, Figs. 154-155). Some abscesses may even perforate the bone (*e.g.* Clutton-Brock *et al.* 1990, 10, Pl.3) as has been found on archaeological specimens of sheep (Fig. 2.3) and pig (von den Driesch 1972, Taf. 15/Abb. 55).

**Figure 2.3**: ruptured abscess in the buccal area between the first molar and (broken) third premolar on a mandible of a Copper Age sheep from Horum Höyük, Turkey, caused by a root infection of the fourth premolar (lost *in vivo*). Scale bar: 10 mm.

A special form of infection in ruminants, actinomycosis mandibulae, is a chronic inflammation caused by the bacteria *Actinomyces bovis*, *A. israeli* and *A. lignèresi*. The osseous tissue of the infected mandible becomes 'lumpy', swollen and spongy (*spina ventosa*), then deteriorates (Tamás 1987, 51, Fig. 55). However, only two known cases have been recorded in archaeological assemblages (Siegel 1976; Teichert 1988, 189, 19/7). The latter specimen, a tenth-twelfth century AD find from Dominsel Brandenburg/Havel (former GDR) fell within the geographical region under discussion here.

| | Sheep | Goat | Caprine | Total |
|---|---|---|---|---|
| Cervical vertebra | | | 1 | 1 |
| Scapula | | 4 | | 4 |
| Humerus | | 1 | 1 | 2 |
| Radius | 3 | 1 | 3 | 7 |
| Ulna | | | 4 | 4 |
| Metacarpus | 6 | 3 | | 9 |
| Proximal phalanx | 3 | 3 | 4 | 10 |
| Pelvis (acetabulum) | | | 1 | 1 |
| Calcaneum | | 1 | | 1 |
| Metatarsus | 2 | | 1 | 3 |
| **Total** | **14** | **13** | **15** | **42** |

**Table 2.3**: summary of the taxonomic/anatomical distribution of arthritic symptoms in the material.

## Arthritis

Arthritic deformations are similarly rare in sheep and goat: a remarkably small number of such cases were found in caprines within the consulted literature (Tab. 2.3), making the comprehensive study of the problem rather difficult.

Although the distal extremity segments (especially the distal ends of metapodia and proximal phalanges) seem to be worst effected (Fig. 2.4), such deformations are also apparent on bones of the front limb (especially the elbow joint; Fig. 2.5), which in domestic ungulates carry, on average, two-thirds of the body weight. It is only in the case of draught animals where this natural balance is disturbed since dynamic loading increases on the hind limbs, demonstrably increasing the incidence of pathological lesions in the hip and hock joints (Bartosiewicz 2006). At least some of the arthritic deformations (*arthritis et periarthritis*; von den Driesch and Boessneck 1969, 49) observed on the phalanges of cattle may be attributed to conditions of keeping and working. Draught exploitation, however, is a most unlikely cause of arthritis in caprines. While exostoses, especially on phalanges, may in part be caused by the ossification of ligaments and tendons in response to strain and tension (Murray 1936, 72; Weidenreich 1924, 34; Weinman and Sicher 1955), hypoperfusion is known to play a major role in the formation of these features as well (Boosman *et al.* 1989, 153), and this is most commonly related to age. However, the regressive arthrosis that develops with old age cannot be distinguished on a purely osteo-morphological basis from arthritis resulting from inflammation (Regöly-Mérei 1962, 139). This difficulty is rarely mentioned in the archaeozoological literature (*e.g.* van Wijngaarden-Bakker and Krauwer 1979, 37). The complexity of the problem is well illustrated by the dozen categories into which modern human medicine classifies rheumatic diseases, many of which are accompanied by "arthritic" skeletal symptoms (Rodnan 1973, 678-679).

## Regional comparison

Having outlined some basic anatomical features of the two major types of conditions represented in the material under discussion here, 92 pathological caprine bones were compared between the two regions (Fig. 2.6). Percentages along the axes of this graph show the proportion of specimens within the regional and pathological groups respectively.

Not only is the number of pathological bone specimens greater in absolute terms in the sample from Central Europe (53) but, as may be estimated from Tab. 2.1, these bones were recovered from some 74,500 caprine remains (0.07% pathological). Near Eastern sheep and goat, on the other hand, were represented by *c.*124,000 specimens (0.03% pathological). While the two-fold difference in these percentages cannot be tested statistically, the relationship between the two types of disease was more directly comparable.

A Chi$^2$ =9.288 (df=1) value calculated between the region and main type of disease as summarised in Fig. 2.6 reveals that the geographical distributions of periodontal disease and arthritis are inverted, and this trend is

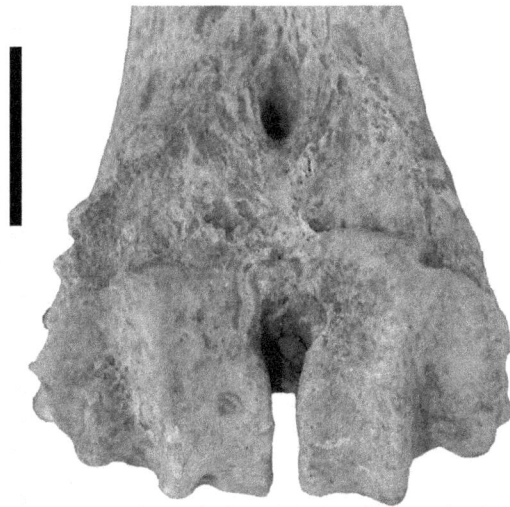

**Figure 2.4**: exostoses on the distal end of an Early Neolithic sheep metatarsus from Ecsegfalva 23, Hungary; plantar aspect. Scale bar: 10 mm.

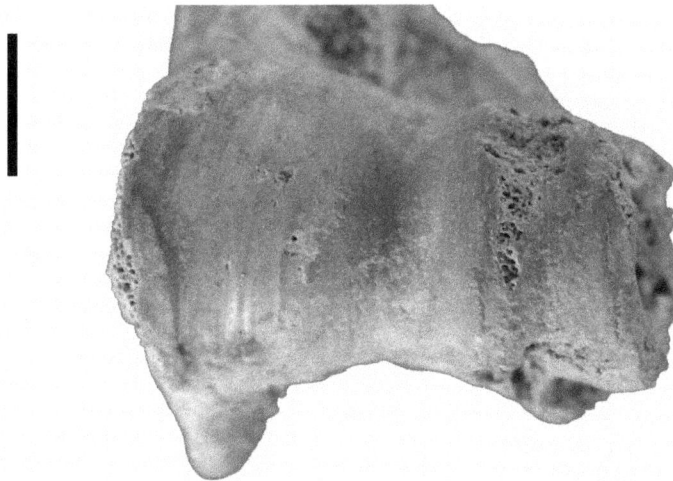

**Figure 2.5**: arthritic grooving on a distal humerus from a Copper Age goat from Horum Höyük, Turkey; anterior aspect. Scale bar: 10 mm.

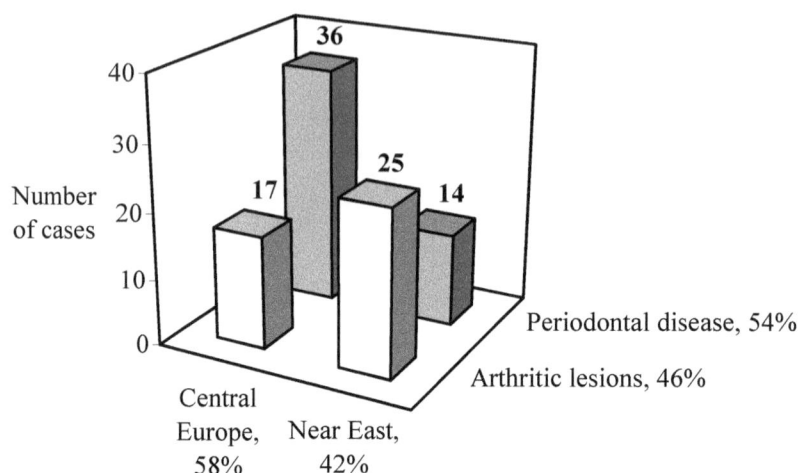

**Figure 2.6**: frequency of 92 pathological caprine bones by geographical region and main type of lesion

statistically significant at the P≤0.01 level of probability: periodontal disease occurs more commonly in caprine bone assemblages in Europe, while arthritic lesions are more frequent in the Near East. The low coefficient of correlation (Phi=0.317) obtained between region and type of disease is indicative of a small but definite relationship. The fact that periodontal disease occurs more commonly in Central Europe, while arthritic disorders are somewhat more characteristic in the Near East is diametrically opposite of what would have been expected on the basis of the stereotypical interpretations of these two types of disease.

## Discussion

On the basis of the material selected for study, the morbidity of sheep and goat was greater in Central Europe than in the Near East. This, in itself, suggests that caprines fared better in their native regions. An underlying issue, possibly influencing the results is that these two species react differently to environmental stress.

The significantly greater frequency of periodontal disease observed at European sites falls in line with the greater general morbidity of caprines in this region. Haimovici and Haimovici (1971, 261-266) concluded that this condition in domesticates increased with 'civilisation' resulting from a combination of environmental features. The general impact of the graze is not limited to the mechanical effects of thorny plants: a variety of infections may follow as a general consequence of deteriorating health. Even in presumably rich pastures, calculus formation may contribute to oral pathologies. Plotting Levitan's aforementioned raw data (Levitan 1977) from Anglo-Saxon England yielded a non-linear relationship characterised by a very high coefficient of determination indicating over 93% ($R^2$ = 0.932, P≤0.001) between calculus formation and the incidence of periodontal disease (Fig. 2.7). With the advancement of calculus build-up, there is a progressive trend in the emergence of

periodontal inflammations. Abscesses also occur most commonly in the third premolar to the second molar region of the mandible (Fig. 2.3). No such systematic data were found regarding the upper tooth row, but the molar area seems to be affected in sheep (Meadow 1983, 392).

Interestingly, a diachronically decreasing incidence of dental lesions was observed in sheep and goat between the pre-Halafian and Early Halafian deposits of the Late Neolithic settlement at Tell Sabi Abyad in Syria (Cavallo 1995, 49; 1997, 59, Tab. 6.14 and 64, Tab. 20). Evidently, trends in the emergence of periodontal disease should by no means be imagined in a straightforward, linear fashion.

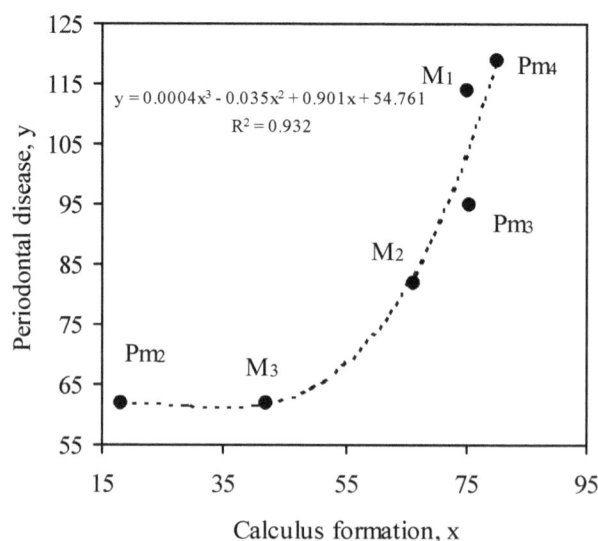

**Figure 2.7**: relationship between the frequencies of calculus and periodontal disease in Anglo-Saxon sheep mandibles from Wicken Bonhunt, England (raw data from Levitan 1977).

The multi-causal aetiology of this condition includes overgrazing. The ridged, swollen roots of teeth in prehistoric caprines from the Near East (*e.g.* Meadow 1983, 392; Grigson 1987, 225) were also interpreted as being indicative of the over-use of infected pastures and

8

crowded keeping conditions. Recent tooth enamel micro-wear studies in Turkey (Beuls *et al.* 2002) have shown that sheep, as opposed to goat, overwhelmingly prefer grasses to shrubs with the opposite tendencies for goat. Sheep switch to herbs in spring but goats remain fixated on the tough shrubs. The greater importance of goat exploitation in the Near East may mean that sheep, more susceptible to periodontal disease were relatively under-represented at several sites. On the other hand, recent dental microwear studies at the Early Neolithic site of Ecsegfalva in Hungary have shown high levels of stocking (Mainland 2007); an idea consistent with the hypothesised environmental stress that sheep and goat may have been exposed to in this region.

The apparent dominance of arthritic deformations in the material from the Near East must be treated with extreme caution. First of all, the greater number of cases used in the calculation originates from a larger sample of caprine bones. More importantly, the anatomical patterning observed in the distribution of arthritic symptoms (Tab. 2.3) is a reminder that many such lesions may result from senescence. On the other hand, at least seasonally poor graze in semi-arid or even arid areas of the Near East means, that the animals must cover greater distances in search of the same quantity of food than in Europe. This may contribute to arthritic deformations, especially in older animals. The condition may also be exacerbated by rough terrain.

## Conclusions

While on the basis of the assemblages under discussion here the morbidity of sheep and goat was greater in Central Europe than in their native areas in the Near East, hypotheses concerning the two major groups of disease did not hold: oral pathologies seem relatively more frequent in Europe in spite of better pastures, while arthritic deformations occurred less commonly than in the Near East. Relating these differences to environmental stress is of special interest. Such cases, however, should be evaluated individually, within the context of each site and period.

One of the main environmental factors in animal keeping is human interference. In the case under discussion, it resulted in the exportation of sheep and goat to less favourable habitats in Europe.

From a cultural point of view, another form of manipulation is worth considering. Pooling several caprine bone assemblages with possibly differing age structures may introduce several arthritic bone specimens in the sample that have less to do with the natural environment than with culling at an older age. In extreme cases, secondary exploitation (*sensu* Sherratt 1983) may itself act as an anthropogenic environmental factor that promotes the incidence of age-related arthritis through an artificially maintained demographic structure. Longevity also increases the statistical probability of trauma, of which, compound fractures often culminate in an arthritic condition during the healing process (Fig. 2.8).

A greater number of studies with special regard to the points raised in this paper will help better understanding

ancient pastoralism and its effect on the health of animals in various natural and cultural environments.

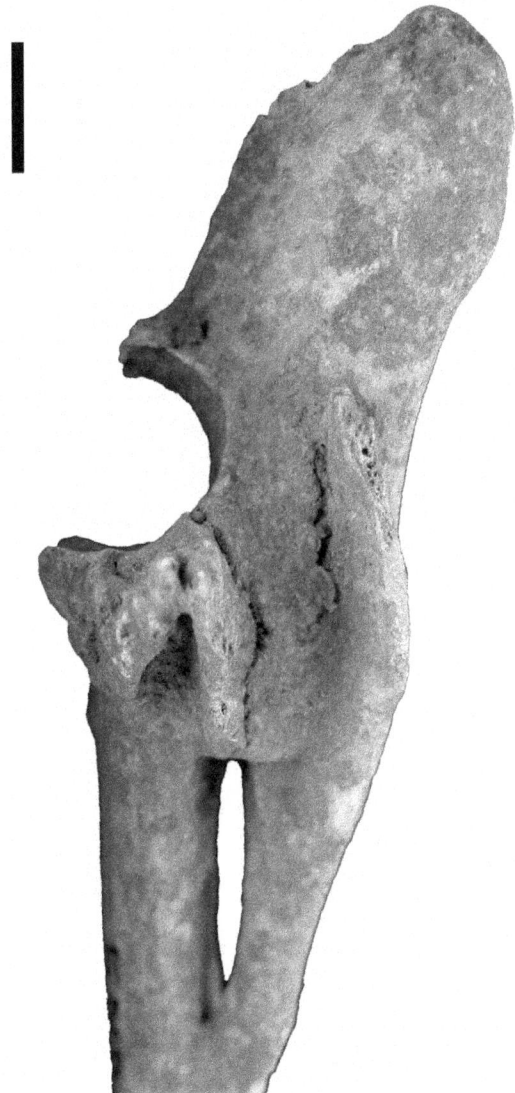

Figure 2.8: arthritic deformation of the ulna related to a compound fracture from an Early Neolithic sheep from Endrőd 119, Hungary; lateral aspect. Scale bar: 10 mm.

## Acknowledgements

This research was carried out within the framework of the project entitled "Anatomical and palaeopathological investigations on the skeletal system of domestic animals" supported by Grant No. T047228 of the National Scientific Research Foundation of Hungary (OTKA). Special thanks are due to Dr. Alice M. Choyke and Dr. Erika Gál, who contributed photographs to this study.

## Bibliography

Anschütz, K. 1966. *Die Tierknochenfunde aus der mittel-alterlichen Siedlung Ulm-Weinhof.* Stuttgart:

Naturwissenschaftliche Untersuchungen für Vor und Frühgeschichte, Württemberg und Hohenzollern 2.

Arbinger-Vogt, H. 1978. *Vorgeschichtliche Tierknochenfunde aus Breisach am Rhein*. Unpublished dissertation. München: Institut für Paläoanatomie, Domesti-kationsforschung und Geschichte der Tiermedizin der Universität München.

Baker, J. R. and Brothwell, D. 1980. *Animal Diseases in Archaeology*. London: Academic Press.

Bartosiewicz, L. 1984. Az állatállomány faji összetételének összefüggése a lakosság étrendjének energiatartalmával néhány fejlõdõ országban. *Állattenyésztés és Takarmá-nyozás* 33/3, 193-203.

Bartosiewicz, L. 1998. Interim report on the Bronze Age animal bones from Arslantepe (Malatya, Anatolia), pp. 221-232, in Buitenhuis, H., Bartosiewicz, L. and Choyke, A. M. (eds), *Archaeozoology of the Near East III*. Groningen: ARC Publicaties 18.

Bartosiewicz, L. 1999. The role of sheep versus goat in meat consumption at archaeological sites, pp. 47-60, in Bartosiewicz, L. and Greenfield, H. (eds), *Transhumant Pastoralism in Southern Europe*. Budapest: Archaeolingua Kiadó.

Bartosiewicz, L. 2002. Pathological lesions on prehistoric animal remains from Southwest Asia, pp. 320-336, in Buitenhuis, H., Mashkour, M., Choyke, A. M. and Al-Shiyab, A. H. (eds), *Archaeozoology of the Near East V*. Groningen: ARC Publicaties 62.

Bartosiewicz, L. 2005. Plain talk: animals, environment and culture in the Neolithic of the Carpathian Basin and adjacent areas, pp. 51-63, in Bailey D. and Whittle A. (eds), *(Un)Settling the Neolithic*. Oxford: Oxbow Books.

Bartosiewicz, L. 2006. Mettre le chariot devant le boeuf. Anomalies ostéologiques liées à l'utilisation des boeuf pour la traction, pp. 259-267, in Pétrequin P., Arbogast, R.-M., Péterquin, A.-M., Van Willigen, S. and Bailly, M. (eds), *Premiers Chariots, Premiers Araires. La Diffusion de la Traction Animale en Europe Pendant les IV^e et III^e Millénaires avant Notre Ère*. Paris: CNRS Editions, CRA Monographies 29.

Bartosiewicz, L. and Choyke, A. M. 1985. Animal exploitation at the site of Csabdi–Télizöldes. *Béri Balogh Ádám Múzeum Évkönyve* 1985, 181-194.

Beuls, I., Vanhecke, L., De Cupere, B., Vermoere, M., Van Neer, W. and Waelkens, M. 2002. The predictive value of dental microwear in the assessment of caprine diet, pp. 337-355, in Buitenhuis, H., Mashkour, M., Choyke A. M. and Al-Shiyab, A. H. (eds), *Archaeozoology of the Near East V*. Groningen: ARC Publicaties 62.

Boosman, R., Németh, F., Gruys, E. and Klarenbeek, A. 1989. Arteriographical and pathological changes in chronic laminitis in dairy cattle. *Veterinary Quarterly* 11/3, 144-155.

Bökönyi, S. 1992. The early Neolithic vertebrate fauna of Endrőd 119, pp. 195-299, in Bökönyi, S. (ed.), *Cultural and Landscape Changes in Southeast Hungary, I. Reports on the Gyomaendrőd Project 1*. Budapest: Archaeolingua Kiadó.

Bökönyi, S. and Bartosiewicz, L. 1998. Tierknochenfunde, pp. 385-424, in Hiller, S. and Nikolov, V. (eds), *Karanovo. Die Ausgrabungen im Südsektor 1984-1992*. Wien: Verlag Ferdinand Berger & Söhne Ges. m. b. h.

Bökönyi, S. and Bartosiewicz, L. 2000. A review of animal remains from Shahr-i Sokhta (Eastern Iran), pp. 116-152, in Mashkour, M., Choyke, A. M. and Buitenhuis, H. (eds), *Archaeozoology of the Near East IVB*. Groningen: ARC Publicaties 32.

Cavallo, C. 1995. Some observations on the animal remains from the pre-Halaf levels of Tell Sabi Abyad, Northern Syria, pp. 45-51, in Buitenhuis, H. and Uerpmann, H.-P. (eds), *Archaeozoology of the Near East II*. Leiden: Backhuys Publishers.

Cavallo, C. 1997. *Animals in the Steppe. A Zooarchaeological Analysis of Later Neolithic Tell Sabi Abyad, Syria*. Unpublished Ph.D. thesis. Amsterdam: Universiteit van Amsterdam.

Clutton-Brock, J., Dennis-Bryan, K., Armitage, P. L. and Jewell, P. 1990. Osteology of the Soay sheep. *Bulletin of the British Museum of Natural History (Zoology)* 56/1, 1-56.

Chaplin, R. E. 1971. *The Study of Animal Bones from Archaeological Sites*. London and New York: Seminar Press.

Davis, S. 1996. Measurements of a group of adult female Shetland sheep skeletons from a single flock: a baseline for zooarchaeologists. *Journal of Archaeological Science* 23, 593-612.

Donnerbauer, H. J. 1968. *Tierknochenfunde aus der Siedlung "Am Hetelberg" bei Gielde/Niedersachsen. II. Die Wiederkäuer*. Unpublished dissertation. München: Institut für Paläoanatomie, Domestika-tionsforschung und Geschichte der Tiermedizin der Universität München.

Driesch, A. von den, 1972. *Osteoarchäologische Untersuchungen auf der Iberischen Halbinsel*. München: Studien über frühe Tierknochenfunde von der Iberischen Halbinsel 3.

Driesch, A. von den, 1975. Die Bewertung pathologisch-anatomischer Veränderungen an vor- und frühgeschichtlichen Tierknochen, pp. 413-425, in Clason, A. T. (ed.), *Archaeozoological Studies*. Amsterdam: North Holland Publishing Company.

Driesch, A. von den, and Boessneck, J. 1969. *Die Fauna des "Cabezo Redondo" bei Villena (Prov. Alicante)*. München: Studien über frühe Tierknochenfunde von der Iberischen Halbinsel 1, 45-106.

Geringer, J. 1967. *Tierknochenfunde von der Heuneburg, einem frühkeltischen Herrensitz bei Hundersingen an der Donau (Grabungen 1959 und 1963). Die Paarhufer Ohne die Bovini*. Stuttgart: Naturwissenschaftliche Untersuchungen zur Vor- und Frühgeschichte in Württemberg und Hohenzollern 5.

Gerlach, R. 1967. *Tierknochenfunde von der Heuneburg einem frühkeltischen Herrensitz an der Donau. Die Wiederkäuer*. Stuttgart: Naturwissenschaftliche Untersuchungen zur Vor- und Frühgeschichte in Württemberg und Hohenzollern 7.

Grigson, C. 1987. Shiqmim: pastoralism and other aspects of animal management in the Chalcolithic of the Northern Negev, pp. 219-241, in Levy, T. E. (ed.), *Shiqmim I. Studies Concerning Chalcolithic Societies in the Northern Negev Desert, Israel 1982-84*. Oxford: British Archaeological Reports International Series 356.

Haimovici, A. and Haimovici, S. 1971. Sur la presence de parodontopathies marginales sur des restes subfossiles de mammifères de stations pre- et protohistoriques du territoire de la Roumanie. *Bulletin de l'Groupe International des Rècherches Stomatologiques* 14, 259-271.

Hanschke, G. 1970. *Die Tierknochenfnde aus der Wüstung Wülfingen II. Die Wiederkäuer.* Unpublished dissertation. München: Institut für Paläo-anatomie, Domestikationsforschung und Geschichte der Tiermedizin der Universität München.

Hescheler, K. and Rüeger, J. 1942. Die Reste der Haustiere aus den neolitischen Pfahlbaudörfern Egolzwil 2 (Wauwilersee, Kt. Luzern) und Seematte-Gelfingen (Baldeggersee, Kt. Luzern). *Vierteljahresschrift der Natrforschende Gesellschaft, Zürich* 87, 383-486.

Holmberg, T. and Reiland, S. 1984. The influence of age, breed, rearing intensity and exercise on the incidence of spavin in Swedish dairy cattle. A clinical and morphological investigation. *Acta Veterinaria Scandinavica* 25/1, 113-127.

Johansson, F. 1987. *Zoologische und kulturgeschichliche Untersuchungen an der Tierresten aus der Römischen Palastvilla in Bad Kreuznach.* Kiel: Schriften aus der Archäologisch-Zoologischen Arbeitsgruppe Schleswig-Kiel 11.

Karrer, H-J. 1982. *Die Tierknochenfunde aus dem latènezeitlichen Oppidum von Altenburg-Rheinau.* Unpublished dissertation. München: Institut für Paläo-anatomie, Domestikationsforschung und Geschich-te der Tiermedizin der Universität München.

Klumpp, G. 1967. *Die Tierknochenfunde aus der mittelalterlichen Burgruine Niederrealta Gemeinde Cazis/Graubünden.* Unpublished dissertation. München: Institut für Paläoanatomie, Domesti-kationsforschung und Geschichte der Tiermedizin der Universität München.

Kokabi, M. 1982. *Arae Flaviae II. Viehhaltung und Jagd im Römischen Rottweil.* Stuttgart: Forschungen und Berichte Vor- und Frühgeschichte Baden-Württemberg 13.

Kussinger, S. 1988. *Tierknochenfunde vom Lidar Höyük in Südostanatolien (Grabungen 1979-86).* Unpublished dissertation. München: Institut für Paläoanatomie, Domestikationsforschung und Geschichte der Tiermedizin der Universität München.

Levitan, B. M. 1977. *Pathological Anomalies in Sheep Mandibles: a Methodological Approach.* Unpublished BSc. Thesis. London: University of London.

Mainland, I. L. 2007. A microwear analysis of selected sheep and goat mandibles, 343-348, in Whittle, A. (ed.), *The Early Neolithic on the Great Hungarian Plain: Investigations of the Körös Culture Site of Ecsegfalva 23, County Békés.* Budapest: Varia Archaeologica Hungarica 21.

Meadow, R. 1983. The vertebrate faunal remains from Hasanlu Period X at Hajji Firuz, pp. 369-422, in Voigt, M. (ed.), *Hasanlu Excavation Reports Volume I: Hajji Firuz Tepe, Iran: the Neolithic Settlement.* Philadelphia: University Monograph of the University Museum, Philadelphia 50.

Murray, P. D. F. 1936. *Bones: a Study of the Development and Structure of the Vertebrate Skeleton.* Cambridge: Cambridge University Press.

Nieberle, K. and Cohrs, P. 1962. *Lehrbuch der speziellen pathologischen Anatomie der Haustiere.* Stuttgart: Paul Parey Verlag.

Piehler, W. 1976. *Die Knochenfunde aus dem spätrömischen Kastell Vemania.* Unpublished dissertation. München: Institut für Paläoanatomie, Domestikation-sforschung und Geschichte der Tiermedizin der Universität München.

Pollok, K. 1976. *Untersuchungen an Schädeln von Schafen und Ziegen aus den frühmittelalterlichen Siedlung Haithabu.* Kiel: Schrifte der archäozoologische Arbeitsgruppe Schleswig-Kiel 1.

Pölloth, K. 1959. *Die Schafe und Ziege des Latène-Oppidums Manching.* München: Studien an vor- und frühgeschichtliche Tierreste Bayerns VI.

Pudek, N. 1980. Untersuchungen an Tierknochen des 13. bis 20. Jahrhunderts aus dem Heiligen-Geist-Hospital in Lübeck. *Lübecker Schrifte der Archäologie und Kulturgeschichte* 2, 107-202.

Rauh, H. 1981. *Knochenfunde von Säugetieren aus dem Demircihüyük (Nordwestanatolien).* Unpublished dissertation. München: Institut für Paläoanatomie, Domesti-kationsforschung und Geschichte der Tiermedizin der Universität München.

Regöly-Mérei, Gy. 1962. *Palaeopathologia II.* Budapest: Medicina Könyvkiadó.

Reichstein, H. 1991. *Die Fauna des germanischen Dorfes Feddersen Wierde 1-2.* Stuttgart: Franz Steiner Verlag.

Rodnan, G. P. (ed.) 1973. Primer on rheumatic diseases. Section 4: Classification of rheumatic disease. *The Journal of the American Medical Association* 224, Suppl. 5, 678-679.

Schatz, H. 1963. *Die Tierknochenfunde aus einer Mittelalterlichen Siedlung Württembergs.* Unpublished dissertation. München: Institut für Paläoanatomie, Domestikationsforschung und Geschichte der Tier-medizin der Universität München.

Schmidt-Pauly, I. 1980. *Römerzeitliche und Mittel-alterliche Tierknochenfunde aus Breisach im Breisgau.* Unpublished dissertation. München: Institut für Paläo-anatomie, Domestikationsforschung und Geschich-te der Tiermedizin der Universität München.

Schwarz, W. 1989. *Tierknochenfunde aus dem Gelände einer Herberge in der Colonia Ulpia Traiana bei Xanten am Niederrhein.* Unpublished dissertation. München: Institut für Paläoanatomie, Domestikationsfor-schung und Geschichte der Tiermedizin der Universität München.

Sherratt, A. 1983. The secondary exploitation of animals in the Old World. *World Archaeology* 15/1, 90-104.

Siegel, J. 1976. Animal palaeopathology: possibilities and problems. *Journal of Archaeological Science* 3, 349-384.

Stahl, U. 1989. *Tierknochenfunde vom Hassek Höyük (Südostanatolien).* Unpublished dissertation. München: Institut für Paläoanatomie, Domest-ikations-forschung und Geschichte der Tiermedizin der Universität München.

Steber, M. 1986. *Tierknochenfunde vom Takht-i Suleiman in der Iranischen Provinz Azerbaidjan (Grabungen 1970-1978).* Unpublished dissertation. München: Institut für Paläoanatomie, Domestikations-forschung und Geschichte der Tiermedizin der Universität München.

Tamás, L. (ed.) 1987. *Állatorvosi Sebészet 2.* Budapest: Mezőgazdasági Kiadó.

Teichert, L. 1988. Die Tierknochenfunde von der sla-wischen Burg und Siedlung auf der Dominsel Brandenburg/Havel (Säugetiere, Vögel, Lurche und Muscheln). *Veröffentlichungen des Museums für Ur- und Frühgeschichte Potsdam* 22, 143-219.

Tormay, B. 1887. *A szarvasmarha és tenyésztése.* Budapest: Országos Magyar Gazdasági Egyesület.

Virchow, R. 1895. Knochen von Höhlenbären mit krankhaften Veränderungen. *Zeitschrift für Ethno-logie* 27, 706.

Walcher, H. F. 1978. *Die Tierknochenfunde aus den Burgen auf dem Weinberg in Hitzacker/Elbe und in Dannenberg (Mittelalter). II. Die Wiederkäuer.* Dissertation. München: Institut für Paläoanatomie, Domestikationsforschung und Ges-chichte der Tiermedizin der Universität München.

Weidenreich, F. 1924. Wie kommen funktionelle an-passungen der aussenform des knochenskelettes zustanze? *Paläontologische Zeitschrift* 7, 34.

Weinmann, J. P. and Sicher, H. 1955. *Bone and Bones. Fundamentals of Bone Biology* (second edition). London: H. Kimpton.

Whittle, A., Bartosiewicz, L., Borić, D., Pettitt, P. and Richards, M. 2002. In the beginning: new radio-carbon dates for the early Neolithic in northern Serbia and south-east Hungary. *Antaeus* 25, 1-51.

Whittle, A., Bartosiewicz, L., Borić, D., Pettitt, P. and Richards, M. 2005. New radiocarbon dates for the Early Neolithic in northern Serbia and south-east Hungary: some omissions and corrections. *Antaeus* 28, 347-355.

Wijngaarden-Bakker, L. van, and Krauwer, M. 1979. Animal palaeopathology. Some examples from the Netherlands. *Helinium* XIX, 37-53.

## Author's affiliation

Institute of Archaeological Sciences
Loránd Eötvös University
1088 Budapest, Múzeum körút 4/B
Hungary

# Appendix

Sites reviewed in this study:

*Central Europe* (Germany marked as before re-unification: FRG and GDR respectively)

| | | |
|---|---|---|
| 1. | Altenburg-Rheinau, FRG | (Karrer 1986, 62) |
| 2. | am Hetelberg, FRG | (Donnerbauer 1968, 82) |
| 3. | Bad Kreuznach, FRG | (Johansson 1987, 84, Abb. 18/b) |
| 4. | Breisach im Breisgau, FRG | (Schmidt-Pauly 1980, 97) |
| 5. | Breisach-Hochstetten, FRG | (Arbinger-Vogt 1978, 112) |
| 6. | Breisach-Münsterberg, FRG | (Arbinger-Vogt 1978, 112) |
| 7. | Colonia Ulpia Traiana, FRG | (Schwarz 1989,117) |
| 8. | Diconche, France | (Bökönyi and Bartosiewicz 1998) |
| 9. | Dominsel Brandenburg/Havel, GDR | (Teichert 1988, 189, 19/8) |
| 10. | Ecsegfalva 23/B, Hungary | (Bartosiewicz unpublished) |
| 11. | Egolzwil 2, Switzerland | (Hescheler and Rüeger 1942) |
| 12. | Eketorp, Sweden | (Boessneck and von den Driesch 1979, 100) |
| 13. | Endrőd 119, Hungary | (Bökönyi 1992, 231-232, Fig. 25) |
| 14. | Feddersen Wierde, FRG | (Reichstein 1991, Taf. 9/6-7) |
| 15. | Haithabu, FRG | (Pollok 1976, 70) |
| 16. | Heiligen-Geist-Hospital,Lübeck, FRG | (Pudek 1980, 155) |
| 17. | Heuneburg, FRG | (Gerlach 1967, 52) |
| 18. | Heuneburg, FRG | (Geringer 1967, 44) |
| 19. | Hundersingen an der Donau, FRG | (Gerlach 1967, Taf. 3/4) |
| 20. | Karanovo, Bulgaria | (Bökönyi and Bartosiewicz 1998) |
| 21. | Kastell Vemania, FRG | (Piehler 1976, 101) |
| 22. | Manching, FRG | (Pölloth 1959, Taf. 3/18b) |
| 23. | Niederrealta, Cazis/Graubünden, Switzerland | (Klumpp 1967, Taf. VIII/18) |
| 24. | Rottweil, FRG | (Kokabi 1982, 126) |
| 25. | Ulm-Weinhof, FRG | (Anschütz 1966, 23) |
| 26. | Unterregenbach, FRG | (Schatz 1963, 21) |
| 27. | Weinberg in Hitzacker/Elbe, FRG | (Walcher 1978, 98) |
| 28. | Wülfingen, FRG | (Hanschke 1970, 72, Fig. 13a) |

*Near East*

| | | |
|---|---|---|
| 1. | Arslantepe, C Anatolia, Turkey | (Bartosiewicz 1998) |
| 2. | Demircihüyük, NW Anatolia, Turkey | (Rauh 1981, 46) |
| 3. | Hassek Höyük, SE Anatolia, Turkey | (Stahl 1989, 64) |
| 4. | Horum Höyük, S Anatolia, Turkey | (Bartosiewicz unpublished) |
| 5. | Lidar Höyük, SO Anatolia, Turkey | (Kussinger 1988, 53) |
| 6. | Sahr-i Sokhta, Sistan, Iran | (Bökönyi and Bartosiewicz 2000) |
| 7. | Takht-i Suleiman, Azerbaijan, Iran | (Steber 1986, 23) |
| 8. | Tel Dor, Area D2, Israel | (Bartosiewicz unpublished) |

# 3. A developmental anomaly of prehistoric roe deer dentition from Svodín, Slovakia

Marian Fabiš, Richard Thomas, Václav Páral and Dušan Vondrák

## Abstract

*Archaeological excavation of an Eneolithic settlement in Svodín revealed two incomplete roe deer (*Capreolus capreolus L., 1758) mandibles (*mandibula sinistra et dextra) dated to the same period. Based on the wear stage of the dentition (both aged 13-15 months), the archaeological context, and the presence of the same morphological anomaly, it is likely that the two specimens belong to the same individual. The anomaly is a developmental condition represented by an extra tooth located on the dorsal edge of the diastema in front of the second premolar. The extra teeth have their roots oriented towards the second premolar while the crowns point towards the* pars incisiva *of the mandibles. Atavistic polyodontia/hyperodontia of the deciduous first premolar and permanent first premolar is suggested as the probable diagnosis.*

## Introduction

The term 'oral pathology' covers a wide range of conditions located in the oral cavity (Baker and Brothwell 1980) and relates to all dysfunctions of both soft and hard tissue structures. Besides traumatic alteration of teeth, caries, neoplasia, and a number of other conditions (*e.g.* Davies 2005), developmental anomalies can also negatively affect the functional ability of the teeth and the whole oral cavity. Such dental anomalies can refer to the number, location, orientation, size and shape of teeth (*e.g.* Baker and Brothwell 1980; Miles and Grigson 1990; Šutta *et al.* 1986; Zendulka *et al.* 1987). *Torsio* (the rotation of a tooth around its longitudinal axis)*, deviation* (the rotation of a tooth around its transverse axis) and *diastasis dentium* (abnormally large spaces between neighbouring teeth in a tooth row) are a few examples of these anomalies. Another example is *transposition*, which refers to the relocation of a tooth to an atypical position (Zendulka *et al.* 1987). Malocclusion, (*i.e.* irregular wear of teeth) can also be caused by developmental disorders of jawbones, such as *brachygnathia superior* or *inferior* (Šutta *et al.* 1986).

Dental anomalies often occur in domestic animals (Šutta *et al.* 1986; Zendulka 1987) and in severe cases cause serious difficulties in fodder intake. Such disorders can even result in the slaughtering of an animal to prevent economic loss. Cases of dental anomalies in domestic species are occasionally reported in studies of faunal assemblages from archaeological sites (*e.g.* Maltby 1979; Baker and Brothwell 1980; O'Connor 1984; Kratochvíl 1986; Albarella *et al.* 1993; Thomas 2005; Fabiš 2004). However, anomalies of dentition also occur in wild species (*e.g.* Ratcliffe 1970; Meyer 1977), although reports on such conditions are virtually absent from the archaeozoological literature. This paper seeks to partially address this imbalance, by considering an anomaly identified in two roe deer (*Capreolus capreolus* L., 1758) mandibles from an archaeological site in Slovakia.

## Materials

Between 1971 and 1983, the Archaeological Institute of the Slovak Academy of Sciences in Nitra undertook a series of archaeological excavations at an Eneolithic (fourth to early-third millennium BC) settlement in Svodín (Němejcová-Pavúková 1986). The site, located in south-western Slovakia (Fig. 3.1) produced a large number of animal bones representing settlement debris. Study of the faunal remains is still in progress; however, amongst the material analysed to date, two roe deer mandible fragments exhibiting a dental developmental abnormality were found (Figs. 3.2-3.7).

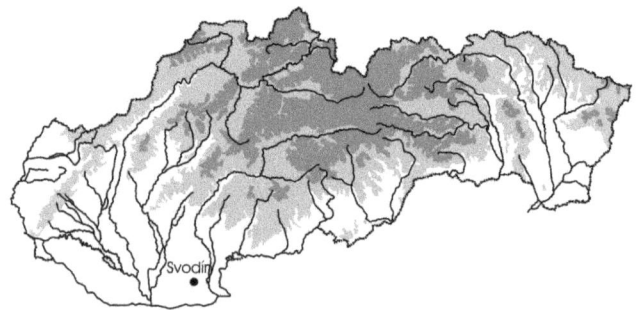

**Figure 3.1**: map of Slovakia with the location of Svodín indicated.

Both jaws (*mandibula dextra*, sample No. 1204 and *mandibula sinistra*, sample No. 1782, found in feature No. 406/77) date to the Lengyel Culture (first half of the fourth millennium B.C). The two samples were located approximately 50 cm from each other. The mandible fragments were analysed on a macroscopic level and radiographically. Estimation of the age of the individual(s) from which they derived was attempted using the eruption and wear of the preserved teeth following the criteria of Habermehl (1961). The samples are currently stored at the private veterinary laboratory of M. Fabiš.

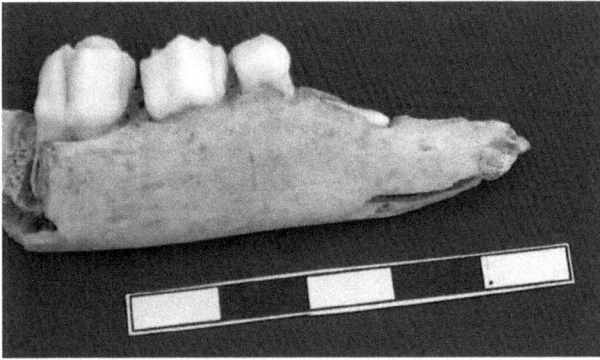

**Figure 3.2**: *mandibula dextra*; lateral aspect.

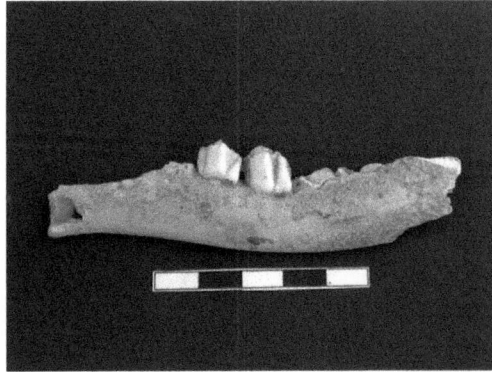

**Figure 3.3**: *mandibula sinistra*; lateral aspect.

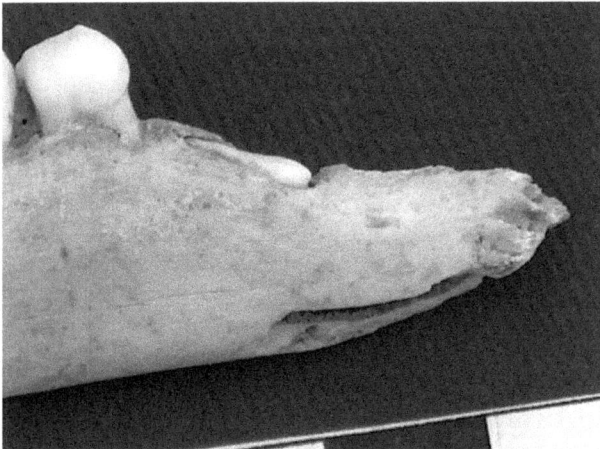

**Figure 3.4**: *mandibula dextra*; detail.

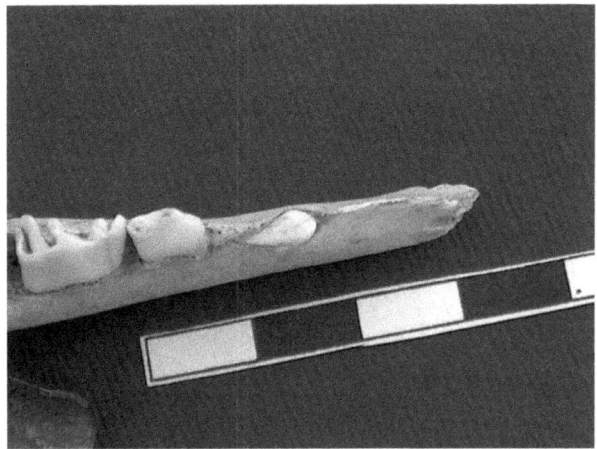

**Figure 3.5**: *mandibula dextra*; dorsal aspect.

**Figure 3.6**: *mandibula sinistra*; detail.

**Figure 3.7**: *mandibula sinistra*; dorsal aspect.

## Results

In both specimens, only parts of the corpus together with a portion of the cheek teeth and diastema are preserved. In the case of the right mandible no other teeth except the premolars (P2, P3, and P4) are present (Fig. 3.2). In the left mandible, the third and fourth premolar are preserved complete (Fig. 3.3). The crowns of the second premolar and first and second molars have been broken off, but their roots are fixed in their alveoli. *Pars incisiva mandibuale* is missing in both samples. The dentition of the two mandibles reveals a similar stage of wear. The premolars are permanent, with the fourth premolar still unworn. Applying the dental ageing criteria of Habermehl (1961), the wear stage indicates that both mandibles belong to individual(s) of 13-15 months age.

Regarding the dentition, an additional tooth closely anterior to the second premolar is present in each of the mandibles. In both cases the tooth is located on the dorsal edge of the diastema and is partially visible from lateral, medial, and dorsal views, although the majority of both teeth is located within the mandibular bone. The visible

15

portion of the tooth is club-shaped with a thin (and sharp) root section oriented towards the second premolar, while the rounded crown section directs towards the missing *pars incisiva* of the jaw. The long axis of the tooth runs parallel to the long axis of the *corpus mandibulae*. This description refers to both samples presented in the paper (Figs. 3.4-3.7).

Radiographic examination of the mandibles confirms the horizontal orientation of the long axis of the teeth with their roots oriented towards the second premolar. The radiograph of the left mandible (Fig. 3.8) shows that the crown of the supernumerary tooth is almost rounded, and the tooth has two closed roots. The radiograph of the right mandible (Fig. 3.9) shows that the crown of the supernumerary tooth is very similar in shape to the extra tooth in the left mandible; however, it has a single root that is open.

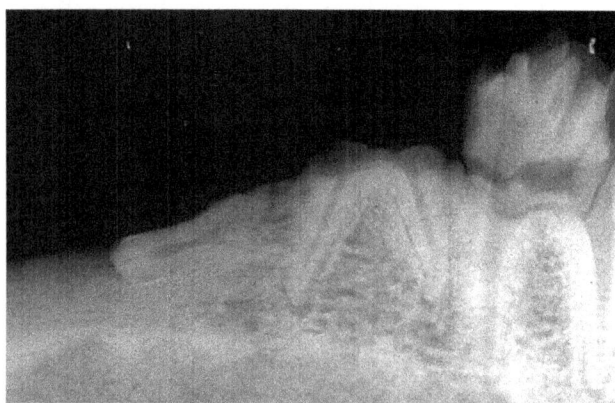

**Figure 3.8**: radiograph of the left mandible.

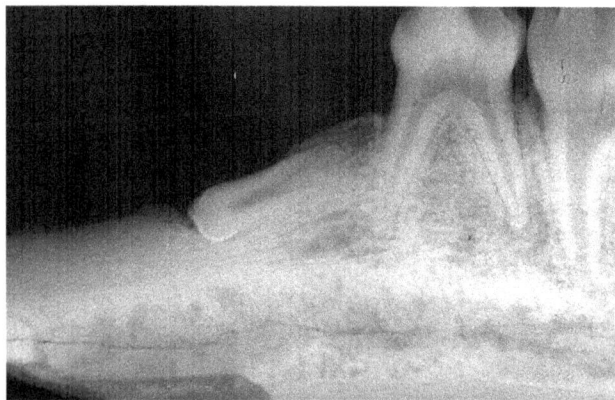

**Figure 3.9**: radiograph of the right mandible.

## Discussion

The first question to address is whether we have got two different roe deer individuals showing the same dental anomaly or whether it is only one individual. Judging from the close spatial location of the specimens, their date, the fact that they are of the same physical age and similarly sized, the two specimens probably belong to the same individual. The fact that both mandibles exhibit the same dental anomaly supports this conclusion.

The tooth formula for roe deer permanent dentition is: 0I, 0C, 3P, 3M / 3I, 1C, 3P, 3M, while for deciduous dentition it is: 0Id, 0Cd, 3Pd / 3Id, 1Cd, 3Pd (Habermehl 1961; Komárek *et al.* 2001). Although the dentition of the two roe deer mandibles from Svodín has not been preserved complete, it clearly does not fit into the roe deer dental formula. It is certain that the described teeth located in front of the second premolar are supernumerary and according to their position they appear to be first premolars.

Our attempts to find any parallel for this condition in the archaeozoological literature failed. In this context it seems that we are dealing with a rare developmental anomaly of roe deer dentition. However, while there are no archaeozoological references of similar conditions reports of extra teeth in cervids have been identified in modern veterinary literature (Miles and Grigson 1990, 110-114). As early as the middle of the last century Wirchov (1940) reported on supernumerary teeth in lower jaws of roe deer. Chaplin and Atkinson (1968) have also reported on the occurrence of the upper canine teeth in roe deer male and female skulls collected from England and Scotland. Jackson (1975) provides results of a roe deer mandibular study in which two specimens out of 51 exhibited an extra tooth located in front of the permanent premolars. Prior (1968) has also noted extra premolars in two roe deer lower jaws, and Meyer (1985) has published an article on the presence of the first premolar in roe deer mandibles. In studies of roe deer from the former Czechoslovakia, the anomalies identified included polyodontia of the first premolar (Kratochvíl 1984; Zima 1988). A further study reported on the occurrence of the deciduous and permanent fist premolar in white-tailed deer (*Odocoileus virginianus*; Mech *et al.* 1970). In addition, a number of animal dentition studies have focused on the ontogeny of dentition in particular species (*e.g.* Kierdorf 1993; Witter and Míšek 1999). Kierdorf's (1993) study deals with dentition ontogeny in roe deer, describing and discussing supernumerary teeth in front of the mandibular second premolar.

Primitive artiodactyls possessed complete dentition (Musil 1987). During the course of evolution, particularly the phylogeny of mammals, quantitative and qualitative changes of dentition have occurred, a process evident in the palaeontological record. Among the changes, a reduction in the number of teeth occurred. For even-toed ungulates (artiodactyls), such as bovids and cervids, the upper incisors and first upper and lower premolars disappeared. The upper canines in bovids also disappeared, but for cervids this is not the rule since they can sporadically occur – as for example in roe deer (*e.g.* Kratochvíl 1984; Zima 1988) but also in some other cervids. Witter and Míšek (1999) studied the development of dentition in sheep foetuses and found that during ontogeny, rudimental primordia of upper incisors, upper canines and lower first premolars appear in the dental lamina. These primordia, however, soon disappear and the teeth do not develop. Similarly, Kierdorf (1993) reported that rudimental primordia of mandibular first premolars appear in roe deer foetuses. Normally, these primordia regress during further growth of the foetus. In the event that the primordia continue their development, first premolars appear in front of the second premolar (*e.g.* Kierdorf 1993, Figs. 3-4). The occurrence of teeth formerly lost in the course of phylogeny is termed atavistic polyodontia and/or hyperodontia (Šutta *et al.* 1986; Zima 1988; Miles and Grigson 1990; Kierdorf 1993). According to Kierdorf (1993), single-rooted teeth with simple conical crowns represent the deciduous first premolar, while teeth with two roots and more complex crown morphology represent the permanent first premolar. As noted above, X-ray examination of the two specimens provides very important information regarding their form (Figs. 3.8 and 3.9). Firstly, the supernumerary tooth of the right mandible has a single root that is not closed, which is typical of deciduous teeth. Secondly, the supernumerary tooth of left mandible has two roots, which are closed, a feature characteristic of permanent teeth. This evidence, together with Kierdorf's criteria, suggest that the Svodín findings represent one or more individuals suffering from atavistic polyodontia having developed its first premolars – the right mandible with a deciduous first premolar and the left mandible with a permanent premolar.

## Conclusions

Two roe deer mandibles found at the Eneolithic settlement in Svodín, which probably derived from the same individual, bear what appears to be an identical anomaly of their dentition; an extra tooth located on the dorsal edge of the diastema in front of the second premolar. Only a small part of the teeth had erupted out of the alveolus, while the remainder stayed buried within the mandible. The teeth are clearly supernumerary and their location and appearance indicate that they are first premolars. The condition is a developmental anomaly diagnosed as atavistic polyodontia/hyperodontia and seems to be the first and only case of this anomaly recognized and reported to date in the archaeozoological literature, at least of central-European origin.

## Acknowledgments

The authors are grateful to all of those who entered the discussion on the supernumerary teeth in roe deer either via web communication and/or e-mail contacts. Our special thanks go to Uwe Kierdorf, Richard Prior and Jan Zima.

## Bibliography

Albarella. U., Ceglia, V. and Roberts, P. 1993. *Giacomo degli Schiavoni (Molise): An Early Fifth Century AD Deposit of Pottery and Animal Bones from Central Adriatic Italy.* Papers of the British School of Rome, Vol. LXI. Oxford: Alden Press.

Baker, J. R. and Brothwell, D. R. 1980. *Animal Diseases in Archaeology.* London: Academic Press.

Chaplin, R. E. and Atkinson, J. 1968. The occurrence of upper canine teeth in roe deer (*Capreolus capreolus*) from England and Scotland. *Journal of Zoology* 155, 141-144.

Davies, J. J. 2005 Oral pathology, nutritional deficiencies and mineral depletion in domesticates – a literature review, pp. 80-88, in Davies, J., Fabiš, M., Mainland, I., Richards, M. and Thomas, R (eds), *Diet and Health in Past Animal Populations: Current Research and Future Directions.* Oxford: Oxbow.

Fabiš, M. 2004. Palaeopathology of findings among archaeofaunal remains of small seminar site in Nitra. *Acta Veterinaria Brno* 73, 55-58.

Habermehl, K. H. 1961. *Alterbestimmung bei Haustieren, Pelztieren und beim Jagdbaren Wild.* Berlin: Paul Parey.

Jackson, J. 1975. Mandibular dental abnormalities in roe deer (*Capreolus capreolus*) from the New Forest. *Journal of Zoology* 177, 491-493.

Kierdorf, von H. 1993. Das Auftreten mandibularer und maxillarer erster Praemolaren beim Reh (*Capreolus capreolus* L.) in ontogenetischer und evolutiver Sicht. *Zoologische Jahrbucher für Anatomie und Odontologie der Tiere* 123, 227-243.

Komárek, V., Štěrba, O. and Fejfar, O. 2001. *Anatomie a Embryologie Volně Žijících Přežvýkavců.* Praha: Grada Publishing.

Kratochvíl, Z. 1984. Veränderungen am Gebiss des Rehwildes (*Capreolus capreolus* L.). *Folia Zoologica* 33 (3), 209-222.

Kratochvíl, Z. 1986. Das Fehlen des 2. Prämolaren beim europäischen Reh (*Capreolus capreolus* L.) aus der jüngeren Steinzeit. *Zeitschrift für Jagdwissenschaft* 32 (4), 248-251.

Maltby, M. 1979. *Faunal Studies on Urban Sites. The Animal Bones from Exeter, 1971- 1975.* Vol. 2. H. Charlesworth, Huddersfield.

Mech, L. D., Frenzel, L. D., Karns, P. D. and Kuehen, D. W. 1970. Mandibular dental anomalies in white tailed deer from Minnesota. *Journal of Mammalogy* 51, 804-806.

Meyer, P. 1977. Angeborne Oligodontien beim Rehwild (*Capreolus capreolus* L.). *Zeitschrift für Jagdwissenschaft* 23, 98-100.

Meyer, P. 1985. Beidseitiger P1 im Unterkiefer eines Rehs (*Capreolus capreolus* L.). *Zeitschrift für Jagdwissenschaft* 31, 120-123.

Miles, A. E. W. and Grigson, C. 1990. *Colyer's Variations and Diseases of the Teeth of Mammals.* Cambridge: Cambridge University Press.

Musil, R. 1987. *Vznik, vývoj a vymíraní savců.* Praha: Academia.

Němejcová-Pavúková, V. 1986. Vorbericht über die Ergebnisse der systematischen Grabung in Svodín in den Jahren 1971-1983. *Slovenská Archeológia* 34, 133-176.

O'Connor, T. P. 1984. *Selected Groups of Bones from Skeldergate and Walmgate.* The Archaeology of York 15/1. London: Council for British Archaeology.

Prior, R. 1968. *The Roe Deer of Cranborne Chase.* Oxford: Oxford University Press.

Ratcliffe, P. R. 1970. The occurrence of the vestigial teeth in badger (*Meles meles*), roe deer (*Capreolus capreolus*) and fox (*Vulpes vulpes*) from the country of Agryll, Scotland. *Journal of Zoology* 162, 521-525.

Šutta, J., Orság, A., Janda, J., Kottman, J., Král, E., Nechvátal, M. and Roztočil, V. 1986. *Veterinárna Chirurgia* (second edition). Bratislava/Praha: Príroda/ SZN.

Thomas, R. 2005. *Animals, Economy and Status: the Integration of Zooarchaeological and Historical Evidence in the Study of Dudley Castle, West Midlands (c. 1100-1750).* Oxford: British Archaeological Reports British Series 392.

Zendulka, M., Škarda, R., Černý, L., Groch, L., Halouzka, R., Holman, J., Kaman, J., Konrád, V., Marcaník, J., Pauer, T. and Pivník, L. 1987. *Patologická anatomie hospodářských zvířat.* Praha: SZN.

Zima, J. 1988. Incidence of dental anomalies in *Capreolus capreolus* from Czechoslovakia. *Folia Zoologica* 37 (2), 129-144.

Witter K. and Míšek I. 1999. Time programme of the early tooth development in the domestic sheep (*Ovis aries*, Ruminantia). *Acta Veterinaria Brno* 68, 3-8.

Wirchov H. 1940. Uberzahlige Wangenzahn im Unterkiefer eines Rehes. *Anatomische Anzeiger* 89, 225-238.

## Authors´ affiliation

Marián Fabiš
Necseyho 17
949 01 Nitra
Slovakia

Richard Thomas
School of Archaeology and Ancient History
University of Leicester
University Road
Leicester LE1 7RH
United Kingdom

Václav Páral
Institute of Anatomy, Histology and Embryology
Faculty of Veterinary Medicine
Veterinary and Pharmaceutical University
Brno
Czech Republic

Dušan Vondrák
Department of Zoology and Anthropology
Faculty of Natural Sciences
University of Constantine the Philosopher
949 01 Nitra
Slovakia

# 4. A possible case of tuberculosis or brucellosis in an Iron Age horse skeleton from Viables Farm, Basingstoke, England

### Robin Bendrey

## Abstract

*This article presents a reanalysis of bone disease in an Iron Age horse (*Equus caballus, *L. 1758) skeleton from Viables Farm, Basingstoke, England. The distribution of infected bone through the skeleton is considered. Possible causes of systemic infection are discussed through comparison with literature on animal and human cases. Tuberculosis and brucellosis are examined as possible diagnoses.*

## Introduction

Excavations between 1974 and 1976 at Viables Farm, Basingstoke (Fig. 4.1) revealed an Iron Age and Romano-British enclosure (Millett and Russell 1982, 1984). A pit within the area of the enclosure (pit 5) produced a double inhumation deposited with animal bones around and beneath the human skeletal remains (Millett and Russell 1982). Inhumation one was a female aged approximately 35-40 years and inhumation two was a female aged approximately 25-30 years (Millett and Russell 1982, 71-2).

The animal remains from pit 5 consisted of: two complete sheep (*Ovies aries*, L. 1758) burials; two horse (*Equus caballus*, L. 1758) burials – one partial, the other largely complete; and two partial cattle (*Bos taurus*, L. 1758) skeletons (Maltby 1982, 75). The mixing and fragmentation of the horse bones have made it difficult to separate the two skeletons with certainty; however, they do seem to represent a near-complete skeleton and a separate collection of lumbar and thoracic vertebrae and some ribs probably deriving from just one animal (Maltby 1982, 78).

Millett and Russell (1984, 54) date the pit to between the third and first century BC; however, Gibson (2004) suggests a later date for the pit, between the first century BC and the first century AD, contemporaneous with the construction and use of the enclosure. A sample of bone from the tibia of the near-complete horse skeleton was submitted for radiocarbon dating by the author to Oxford University Radiocarbon Accelerator Unit. Unfortunately it did not return a date due to poor yield.

## The horse skeleton

All the pathological specimens described below probably derived from the near-complete horse skeleton. This specimen consisted of the skull, vertebral column (atlas to sacrum) and ribs, both hind limbs, and the right fore limb. Major elements missing included all the bones of the left fore limb (scapula to phalanges) and the right scapula. The skeleton is that of a male with a reconstructed withers height of around 10-11 hands

(Maltby 1982, 78). Analysis of crown height measurements of the mandibular teeth, following Levine (1982), indicated an age of seven to nine years at death. There was no evidence for butchery on the skeleton.

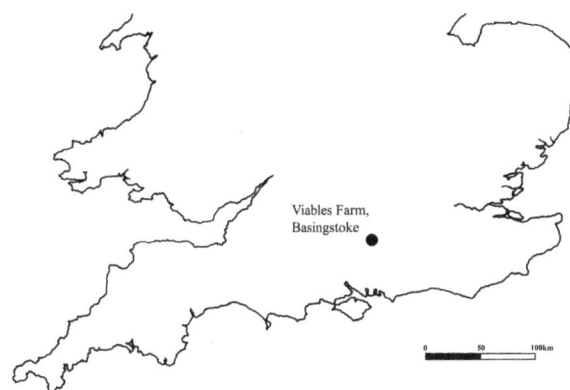

**Figure 4.1**: map showing location of Viables Farm, Basingstoke.

Carnivore gnawing is present on a number of elements of the skeleton, including: the ischium and ilium of the right pelvis; the ischium of the left pelvis; the right distal femur and proximal tibia; the left distal femur; the left calcaneum; the spinous processes of two thoracic and one lumbar vertebra; and the caudal part of the sacrum has been lost due to carnivore gnawing (no caudal vertebrae were present). This evidence points to carnivore gnawing on the exposed extremities of the skeleton (*e.g.* the joints of the hind legs and its rear end), indicating that the animal was accessible for a short while to scavengers, before or after placement in the pit. The pattern of skeletal element representation supports this interpretation: separation of the mandible and scapulae would be expected at an early stage of the natural disarticulation sequence (*e.g.* Hill and Behrensmeyer 1984; Méniel 1992, 58). Delayed burial was also a feature suggested by Millett and Russell (1982, 73) on the basis of disturbance of inhumation one and the accumulations of chalk rubble in the layer above that contained the human and animal bones.

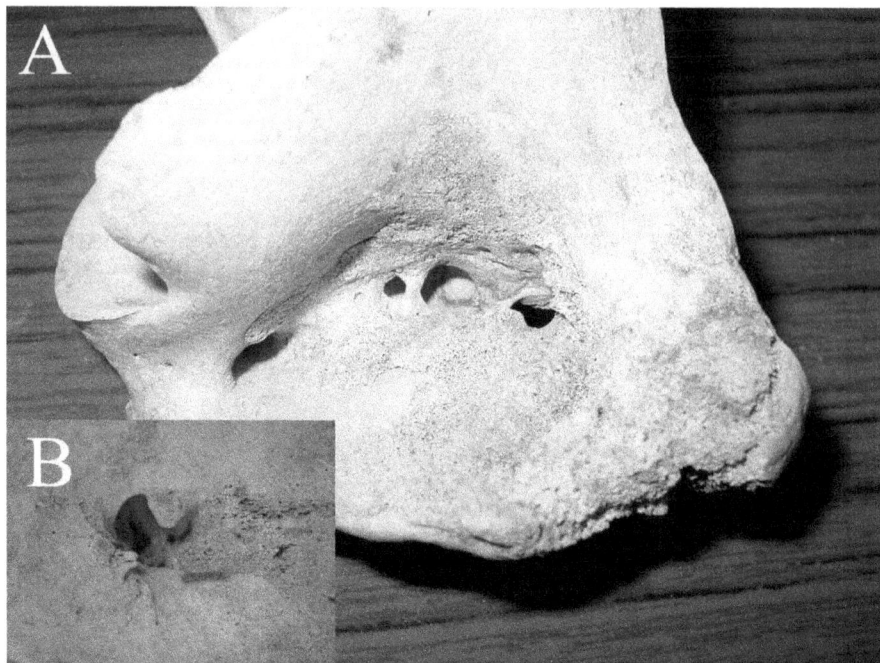

**Figure 4.2**: horse atlas from pit 5, Viables Farm, Hampshire: A – ventral view of atlas showing proliferative periosteal reaction; B – dorsal view of right transverse foramen.

**Figure 4.3**: horse sacrum from pit 5, Viables Farm, Hampshire: A – area of bone proliferation; B – area of possible lysis.

## Description of pathology

Pathology was recorded on the atlas, two thoracic vertebra, the sacrum, four rib fragments and the right pelvis. Proliferative periosteal lesions were identified on the ventral and dorsal surfaces of the right side of the atlas (Fig. 4.2A) and evidence of infection can be seen progressing through the transverse foramen (Fig. 4.2B).

The sacrum also exhibited proliferative periosteal lesions on both the dorsal and ventral surfaces of the left-hand side, with the dorsal surface affected to a greater

affected to a greater degree (Fig. 4.3). In this specimen the infection was seen tracking through the dorsal and ventral sacral foramina and there was also a possible area of bone lysis.

Two thoracic vertebrae revealed proliferative periosteal lesions on the left hand dorsal surface between the transverse process and the spinous process (Figs. 4.4B and 4.5). One of these vertebrae also has destruction of bone around the centrum, in the form of resorptive pits (Fig. 4.4A).

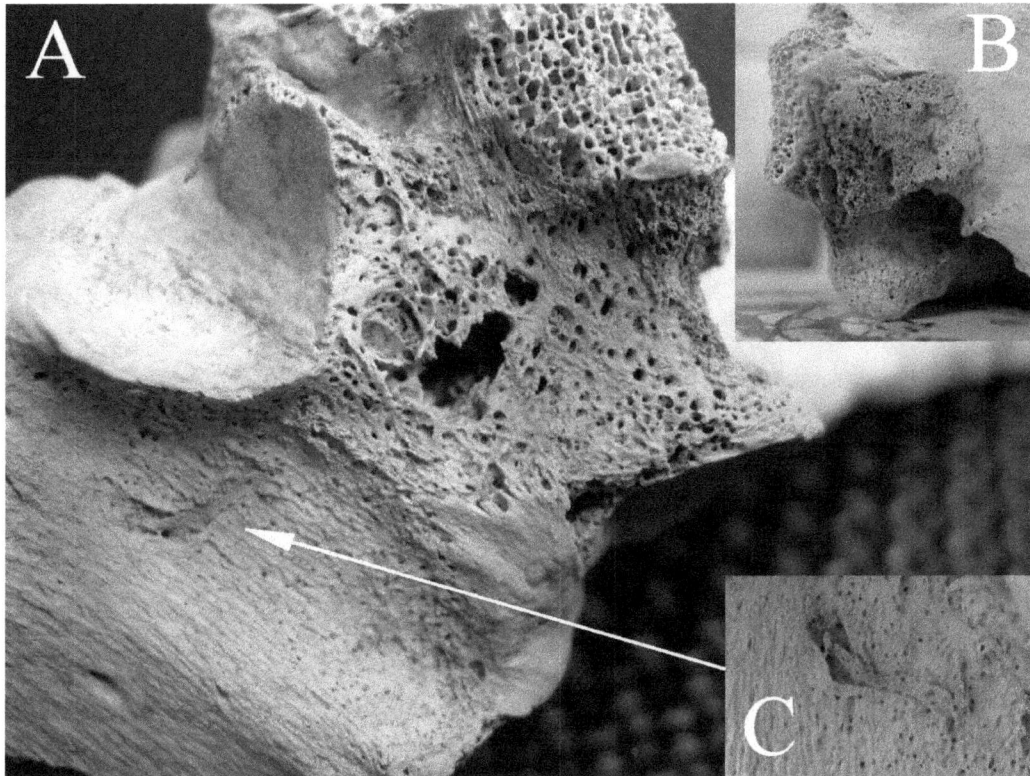

**Figure 4.4**: horse thoracic vertebra from pit 5, Viables Farm, Hampshire: A – destruction of bone around the centrum, in the form of resorptive pits; B – new woven bone above transverse process; C – detail of reactive bone laid down in foramen on left side of centrum.

**Figure 4.5**: horse thoracic vertebra from pit 5, Viables Farm, Hampshire, showing new woven bone above transverse process.

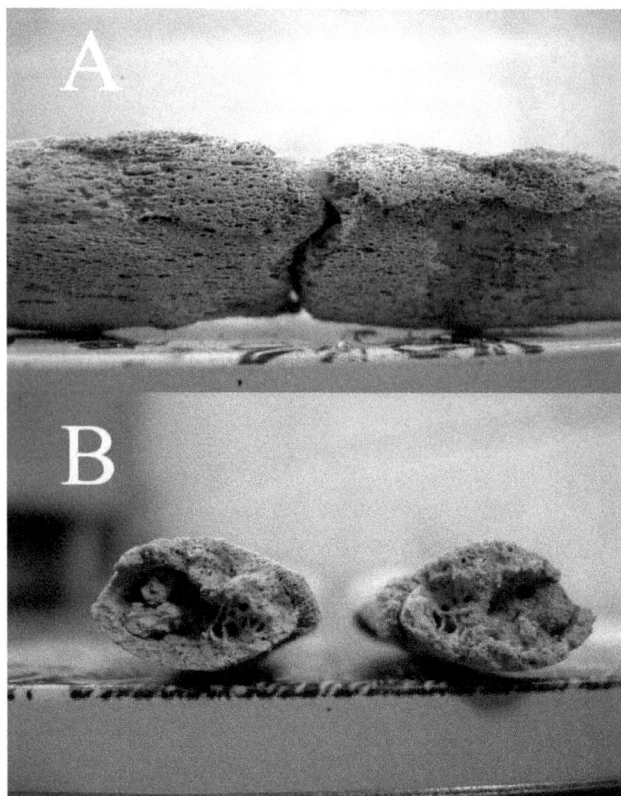

**Figure 4.6**: horse rib from pit 5, Viables Farm, Hampshire: A – proliferative periosteal lesions on the visceral surface; B – destruction of bone causing the rib to weaken and break (post-mortem).

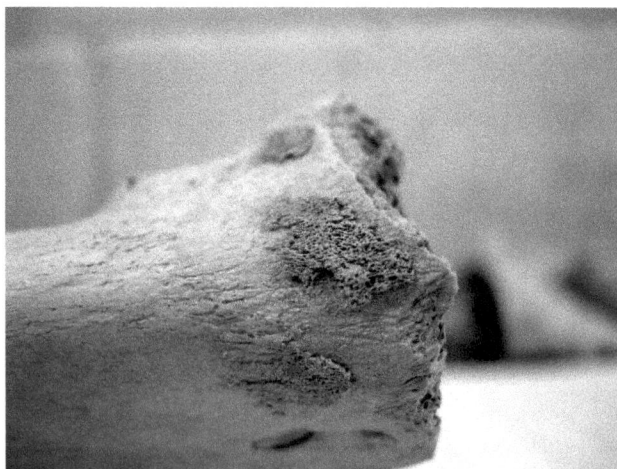

**Figure 4.7**: horse pelvis from pit 5, Viables Farm, Hampshire: dorsal view of right ischial tuberosity showing small areas of new woven bone formation.

Proliferative periosteal lesions were also observed on the visceral surfaces of four rib fragments (*e.g.* Fig. 4.6A). All the rib fragments had broken in the area where the bone had become weakened due to damage caused by the infection (Fig. 4.6B).

A few small areas of new woven bone formation were also recorded on the right ischium (Fig. 4.7).

## Diagnosis

The distribution of infected bones through the skeleton of the Viables Farm horse – on the ribs, vertebrae and pelvis – suggests that the animal was suffering from a systemic infection. Infectious inflammation of bone is usually caused by bacteria, although other agents can be involved (Weisbrode 2007, 1076). A range of bacteria have been implicated in cases of vertebral body osteomyelitis in modern horses (Markel *et al.* 1986, 634), of which *Mycobacterium bovis* (causing tuberculosis) and *Brucella abortus* (causing brucellosis) have been isolated from adult horses (Collins *et al.* 1971; Kelly *et al.* 1972).

There is limited information on the manifestation of tuberculosis in horses. Lignereux and Peters (1999, 344) present information from earlier publications on the disease's manifestation and suggest that granulomatous bone localizations are observed, but they are rare, and that tuberculosis of the spine is located more frequently in the cervical vertebrae. Primary infection in animals is generally respiratory, sometimes digestive and exceptionally through the skin (Lignereux and Peters 1999, 340). The presence of lesions on the visceral surfaces of the ribs suggests the primary infection was respiratory. Cross infection could have occurred from other mammal hosts, perhaps cattle, or from birds. Following infection, and the formation of a primary complex, the next stage corresponds to the spread of the tubercle bacillus in the body by means of the circulatory or lymphatic system (Lignereux and Peters 1999, 340). Generally, in domestic mammals, tuberculosis affects vascular cancellous bone. In cattle, the distribution of tuberculosis lesions in the skeletons "is governed by the usual factors in haematogenous osteomyelitis. The lesions are most frequent in the vertebrae, ribs, and flat bones of the pelvis – all bones that are spongy and highly vascular" (Dungworth 1993, 648). Lignereux and Peters (1999, 341) state that tuberculosis "results in a rarefying osteitis. It begins with osteopenia, *i.e.* demineralisation and osteoporosis, and is followed by an ulcerating or cavitary osteolysis".

As stated, another bacterial infection that may cause such damage to the vertebrae is brucellosis (Collins *et al.* 1971). Aufderheide and Rodríguez-Martín (1998, 192) state that horses (as well as goats, sheep, pigs and cattle) are important reservoirs for human infection, and that cross infections can occur between various animal species. Brucellosis may induce granulomatous bone lesions in horses resulting in cervical or lumbar spondylodiscitis (Lignereux and Peters 1999, 345). Baker and Brothwell (1980, 76) state that the occurrence of brucellosis in the bones of horses, especially the cervical and lumbar vertebral bodies results in erosion, often in a rather irregular fashion, and extensive irregular periosteal new bone proliferation; and it also extends into and destroys the inter-vertebral disc, "an uncommon feature with other forms of vertebral osteomyelitis". Brucellosis lesions resemble those of tuberculosis, although the amount of periosteal proliferation is less in the latter (Baker and Brothwell 1980, 76).

It is also possible that one or more different diseases may be the cause of the noted pathological lesions in the Viables Farm horse, for example with tuberculosis or brucellosis affecting the vertebrae, and an independent pulmonary infection causing the lesions on the visceral surfaces of the ribs.

These diseases in humans have received far greater attention than in horses and a brief look at some of this literature may prove informative for the analysis of the Viables Farm horse. Tuberculosis is contracted by humans through the gastrointestinal tract from infected meat or milk from animals and through the lungs from infected humans and other animals (Roberts 1999, 312). Following infection and establishment in the body the bacteria, spread via the blood and lymphatic system to the skeleton, tends to lodge themselves in the vertebrae primarily (Roberts 1999, 312). In archaeological human remains diagnosis of tuberculosis relies mainly upon the presence of severe vertebral involvement, including collapse and kyphosis; however, less severe lesions must precede these indicators (Baker 1999, 301). Destructive lesions and new bone formation may both occur in the rib cage (Roberts 1999, 312). Although lesions on the internal surfaces of ribs have been correlated with tuberculosis, they are generally considered non-specific indicators of chronic pulmonary infection (Baker 1999, 301). However, proliferative lesions on the visceral surfaces of ribs are frequently accompanied by smooth-walled resorptive pits on the ventral aspects of thoracic and lumbar vertebral bodies; and these lesions often occur in individuals with other pathological indicators of tuberculosis (Baker 1999, 301).

Aufderheide and Rodríguez-Martín (1998, 140, 193) suggest that brucellosis can be separated from tuberculosis, in humans, because brucellosis does not usually cause paravertebral abscesses, vertebral collapse or gibbus formation in vertebrae and it tends to present destructive and reparative processes simultaneously. Conversely, tuberculosis is largely destructive and periosteal proliferation in long bones may be prominent. They do not make any mention of rib involvement by brucellosis in humans, presumably because the organism usually enters the body via the gastrointestinal tract or cuts and abrasions of the skin (Aufderheide and Rodríguez-Martín 1998, 192). However, the involvement of the respiratory system in human cases of brucellosis has been identified as a rare event (Pappas *et al.* 2003), and infection through inhalation of the bacteria could therefore have caused the rib lesions in the Viables Farm horse. McCaughey (1969, 103), however, highlights contamination of food with urine or discharges by animals infected with the *Brucella* organism as the common form of transmission to modern domestic animals. McCaughey also suggests the possibility of transmission by ectoparasites.

Attempting a diagnosis in a species for which there is minimal comparative data is fraught with difficulty. Here, published evidence for tuberculosis and brucellosis in humans has been included in order to broaden the available picture. Cases of vertebral osteomyelitis have been associated with both brucellosis and tuberculosis in adult published horses (Collins *et al.* 1971; Kelly *et al.* 1972); however in the absence of adequate published criteria to separate the two diseases it is not possible to suggest which is most likely from the lesions present in the Viables Farm horse. Although differences between the two diseases are suggested for lesions of the spinal column in humans (*e.g.* Aufderheide and Rodríguez-Martín 1998), whether or not these can be transposed to horses is uncertain as inter-species differences in skeletal manifestation can occur. In dogs, for example, the skeletal involvement of tuberculosis results in bone cell proliferation rather than the demineralisation of bone as seen in humans (Bathurst and Barta 2004, 919).

| | Infection known | Infection unknown |
|---|---|---|
| 'Rubbish' disposal | Discard of diseased animal carcass | Discard of animal carcass |
| 'Ritual' deposition | Economic reasons affecting choice of grave goods (the deliberate slaughter of this diseased animal would have represented a negligible economic loss) | Placement of horse skeleton as grave good, which would have represented a significant economic loss to the community |

Table 4.1: possible interpretations of the nature of the deposit.

The rib lesions on the visceral surfaces of the ribs in the Viables Farm horse suggest a primary pulmonary infection (although it is possible that the infection spread to the lungs at a later stage). The interpretation of the lesions being put forward is as follows and is applicable to both tuberculosis and brucellosis:

1. the new reactive bone formation on the internal surfaces of the ribs indicates a (probable) pulmonary primary infection in the horse (Fig. 4.6);

2. the new reactive bone formation on various vertebrae, from the atlas to the sacrum, and the pelvis indicates the spread of this infection (Figs. 4.2-4.5 and 4.7): it can be seen tracking through the foramina on these bones (Figs. 4.2B and 4.4C);

3. the resorptive pits in the thoracic vertebra (Fig. 4.4A) indicate the establishment of the bacteria in this bone.

The general consensus, from the veterinary and human literature is that tuberculosis is a more common respiratory infection than is brucellosis. This could suggest that the Viables Farm horse was more likely suffering from tuberculosis.

## Discussion

Tuberculosis or brucellosis are suggested as possible causes of the pathological manifestations in the horse skeleton from Viables Farm. This finding is of potential importance for furthering our understanding and identification of these zoonotic diseases in the past. In recent years interest has focussed on the identification of tuberculosis in animal populations, not least because there is an increased awareness of their role in transmitting the disease to human population (Lignereux and Peters 1999; Roberts 2000, 151-2; Mays 2005). In addition, brucellosis in horses has been recognised as a potential source of infection to humans and other animals (Denny 1972).

The domestication of aurochs (wild cattle) has been argued as a major factor that would have probably greatly increased the effect of tuberculosis in both cattle and human populations, with domestic cattle kept in herds and enclosed spaces that enhanced exposure to infected animals (Ortner 1999, 255). Although evidence for tuberculosis in humans is known from Neolithic Europe, the first evidence from Britain is from the middle Iron Age and has been dated to 400-230 BC (Mays and Taylor 2003).

Modern comparative data on tuberculosis in domestic animals is rather limited, due to collective prophylactic measures (systematic tuberculinations and elimination of positive animals) and incineration of carcasses identified as positive at slaughterhouses (Lignereux and Peters 1999, 347). All domestic mammals are susceptible to tuberculosis; however, statistics show that cattle, pigs and carnivores are predominantly affected, compared to the relatively low frequency in horses, sheep and goats (Lignereux and Peters 1999, 347). At the beginning of the twentieth century in Germany, according to Eber (1932, 256, cited in Lignereux and Peters 1999, 344), 0.15 to 0.3% of the horses slaughtered were tuberculous. Francis (1963, 184) states that a prevalence rate of 0.2% has been recorded, but that it is usually much lower than this. The incidence of *Brucella abortus* infection in horse populations indicates that this disease is more common than tuberculosis, with modern surveys giving a range of results from 0.5 to 40% of horses being positive (Denny 1973, 121, table I).

An identification of tuberculosis or brucellosis in the Viables Farm horse has implications in terms of the exposure of the horse to these diseases, and possibly its use, as tuberculosis and brucellosis in horses are both thought to occur principally through contact with cattle (Denny 1973, 121; Mays 2005, 126). Given the low incidence of tuberculosis in horses and the fact that the primary infection is generally respiratory, it is possible that the route of infection to this animal may have come from stabling, or frequent association, with cattle. Brucellosis could also have been contracted in the same way as there is strong modern evidence that cattle are the main source of *Brucella abortus* infection for horses (Denny 1973); although other transmission routes, such as directly from humans, other infected animals, or via the alternate pathways mentioned above are equally plausible.

An identification of tuberculosis or brucellosis in the Viables Farm horse also has implications in terms of the exposure of humans to these diseases. For example, it is known from nineteenth and early twentieth century historical accounts, that tuberculosis in humans and animals was tightly connected (Lignereux and Peters 1999, 341).

The palaeopathological study can also contribute to the interpretation of the combined deposit in pit 5 at Viables Farm. The deliberate deposition of horses is recognised as both complete burials and disarticulated remains at Iron Age sites, and is identified in a range of contexts (*e.g.* Bendrey *et al.* forthcoming; Grant 1984; Legge

1995). The interpretation of complete and partial articulating skeletons in the archaeological record of the Iron Age in southern Britain is a topic that has received much attention, particularly in determining, for example, whether a complete skeleton represents rubbish disposal or 'ritual' deposition (*e.g.* Grant 1984; Hill 1995; Wilson 1992). A number of possible interpretations of this deposit can be suggested based on the zooarchaeological and contextual evidence (Tab. 4.1). The association, at this site, with human inhumations indicates a ritual connotation (*e.g.* Wilson 1992, 341). Interpretation of the inclusion of the horse skeleton in the deposit in part depends on whether the disease was evident in the live animal (Tab. 4.1). At around seven to nine years of age the horse would have been in its prime as a work animal, and killing it to deposit in the pit could represent significant loss to an individual or community. However, the extent of pathological bone in the skeleton suggests that the animal's illness would have been evident to the human owners, and also that it may no longer have been fit for work. This probably influenced the choice of animal to be included in the pit group with the human bodies. As such, inclusion of this horse in the burial deposit would not have represented the loss of a valuable animal; rather, the horse may simply be a representative of its kind – perhaps it was the idea of the horse and what it represented, not the horse itself, that was important (as discussed for dogs by Bendrey *et al.* forthcoming).

## Conclusions

Evidence for bone disease on the ribs, vertebrae and pelvis of the Viables Farm horse skeleton suggests that the animal was suffering from a systemic infection. Following analysis of the distribution of these lesions and consultation with veterinary and medical literature, it is possible to suggest that the horse may have been suffering from tuberculosis or brucellosis. In horses, both diseases are thought to predominantly occur through contact with cattle, which has implications for the ways in which the horse was used and kept. The horse may have also been a source of infection for the attendant human population. The study of the pathology in the horse skeleton has also contributed to interpretations of the inclusion of the animal in a pit with two human inhumations. The identification of zoonotic diseases, such as tuberculosis and brucellosis, in animal populations is particularly important due to the correlation with these diseases in human populations

## Acknowledgements

I would like to thank Kay Ainsworth (Hampshire County Council Museums Service) for allowing me access to the bone assemblage and Stephany Leach and Richard Thomas for commenting on an earlier draft of this paper.

## Bibliography

Aufderheide, A. C., and Rodríguez-Martín, C. 1998. *The Cambridge Encyclopedia of Human Palaeopathology*. Cambridge: Cambridge University Press.

Baker, B. J. 1999. Early manifestations of tuberculosis in the skeleton, pp. 301-307, in Pálfi, G., Dutour. O., Deák, J. and Hutás, I. (eds), *Tuberculosis Past and Present*. Budapest: Golden Book Publisher Ltd/ Tuberculosis Foundation.

Baker, J. and Brothwell, D. R. 1980. *Animal Diseases in Archaeology*. London: Academic Press.

Bathurst, R. R. and Barta, J. L. 2004. Molecular evidence of tuberculosis induced hypertrophic osteopathy in a 16th-century Iroquoian dog. *Journal of Archaeological Science* 31, 917-925.

Bendrey, R., Leach, S. and Clark, K. M. forthcoming. New light on an old rite: reanalysis of an Iron Age burial group from Blewburton Hill, Oxfordshire, in Morris, J. and Maltby, J. M. (eds), *Perceiving Ritual Activity Through Environmental Archaeology*. Oxford: British Archaeological Reports International Series.

Collins, J. D., Kelly, W. R. and Twomey, T. 1971. Brucella-associated vertebral osteomyelitis in a Thoroughbred mare. *The Veterinary Record* 88, 321-326.

Denny, H. R. 1973. A review of brucellosis in the horse. *Equine Veterinary Journal* 5 (3), 121-126.

Dungworth, D. L. 1993. The respiratory system, pp. 539-699, in Jubb, K. V. F., Kennedy, P. C. and Palmer, N. (eds), *Pathology of Domestic Animals Volume 2* (fourth edition). New York: Academic Press.

Eber, A. 1932. Tuberculose des gesflüges; tuberkulose der säugetiere, pp. 217-265, in Stang, V. and Wirth, D. (eds), *Tierheikunde und Tierzucht. Eine Encyclopädie der praktischen Nutztierkunde 10*. Berlin and Vienna: Urban & Schwarzenburg.

Francis, J. 1963, Tuberculosis, pp. 184-185, in Bone, J. F., Catcott, E. J., Gabel, A. A., Johnson, L. E. and Riley, W. F. (eds), *Equine Medicine and Surgery: a Text and Reference Work*. Wheaton, Illinois: American Veterinary Publications.

Gibson, C. 2004, The Iron Age and Roman site of Viables Two (Jays Close), Basingstoke. *Proceedings of the Hampshire Field Club and Archaeological Society* 59, 1-30.

Grant, A. 1984. Survival or Sacrifice? A critical appraisal of animal bones in Britain in the Iron Age, pp. 221-227, in Grigson, C. and Clutton-Brock, J. (eds), *Animals and Archaeology: 4. Husbandry in Europe*. Oxford: British Archaeological Reports International Series 227.

Hill, J. D. 1995. *Ritual and Rubbish in the Iron Age of Wessex*. Oxford: British Archaeological Reports British Series 242.

Hill, A. and Behrensmeyer, A. K. 1984, Disarticulation patterns of some modern east African mammals. *Paleobiology* 10 (3), 366-376.

Kelly, W. R., Collins, J. D., Farrelly B. T., Whitty, B. T. and Rhodes, W. H. 1972. Vertebral osteomyelitis in a horse associated with *Mycobacterium tuberculosis var. bovis (M. bovis)* infection. *Journal of the American Veterinary Radiology Society* 13, 59-69.

Legge, A. J. 1995, A horse burial and other grave offerings, p.146-152, in Parfitt, K. (ed.), *Iron Age Burials from Mill Hill, Deal.* London: British Museum Press.

Levine, M. A. 1982. The use of crown height measurements and eruption-wear sequences to age horse teeth, pp. 223-250, in Wilson, B., Grigson, C. and Payne, S. (eds), *Ageing and Sexing Animal Bones from Archaeological Sites.* Oxford: British Archaeological Reports British Series 109.

Lignereux, Y. and Peters, J. 1999. Elements for the retrospective diagnosis of tuberculosis on animal bones from archaeological sites, pp. 339-348, in Pálfi, G., Dutour. O., Deák, J. and Hutás, I. (eds), *Tuberculosis Past and Present.* Budapest: Golden Book Publisher Ltd/Tuberculosis Foundation.

Maltby, M. 1982. The animal bones, pp. 75-81, in Millett, M. and Russell, D. (eds), An Iron Age and Romano-British site at Viables Farm, Basingstoke. *Proceedings of the Hampshire Field Club Archaeology Society* 40, 49-60.

Markel, M. D., Madigan, J. E., Lichtensteiger, C. A., Large, S. M. and Hornof, W. J. 1986. Vertebral body osteomyelitis in the horse. *Journal of the American Veterinary Medical Association* 188 (6), 632-634.

Mays, S. 2005. Tuberculosis as a zoonotic disease in antiquity, pp. 125-134, in Davies, J., Fabiš, M., Mainland, I., Richards, M. and Thomas, R. (eds), *Diet and Health in Past Animal Populations: Current Research and Future Directions.* Oxford: Oxbow.

Mays, S. and Taylor, G. M. 2003. A first prehistoric case of tuberculosis from Britain. *International Journal of Osteoarchaeology* 13, 189-196.

McCaughey, W. J. 1969. Brucellosis in wildlife, pp. 99-105, in, McDiarmid, A. (ed), *Diseases in Free-Living Wild Animals.* London: Academic Press.

Méniel, P. 1992, *Les sacrifices d'animaux chez les Gaulois.* Paris: Editions Errance.

Millett, M. and Russell, D. 1982. An Iron Age burial from Viables Farm, Basingstoke. *Archaeological Journal* 139, 69-90.

Millett, M. and Russell, D. 1984. An Iron Age and Romano-British site at Viables Farm, Basingstoke. *Proceedings of the Hampshire Field Club Archaeology Society* 40, 49-60.

Ortner, D. J. 1999. Paleopathology: implications for the history and evolution of tuberculosis, pp. 225-261, in Pálfi, G., Dutour. O., Deák, J. and Hutás, I. (eds), *Tuberculosis Past and Present.* Budapest: Golden Book Publisher Ltd/Tuberculosis Foundation.

Pappas, G., Bosilkovski, M., Akritiditis, N., Mastora, M., Krteva, L. and Tsianos, E. 2003. Brucellosis and the respiratory system. *Clinical Infectious Diseases* 37, 95-99.

Roberts, C. A. 1999. Rib lesions and tuberculosis: the current state of play, pp. 311-316, in Pálfi, G., Dutour. O., Deák, J. and Hutás, I. (eds), *Tuberculosis Past and Present.* Budapest: Golden Book Publisher Ltd/Tuberculosis Foundation.

Roberts, C. A. 2000. Infectious disease in biocultural perspective: past, present and future work in Britain, pp. 145-162, in Mays, S. and Cox, M. (eds), *Human Osteology in Archaeology and Forensic Science.* London: Greenwich Medical Media Ltd.

Weisbrode, S. E. 2007. Bone and joints, pp. 1041-1105, in McGavin, M. D. and. Zachary, J. F. (eds), *Pathologic Basis of Veterinary Disease* (fourth edition). St Louis, Missouri: Elsevier.

Wilson, B. 1992. Considerations for the identification of ritual deposits of animal bones in Iron Age pits. *International Journal of Osteoarchaeology* 2 (4), 341-349.

## Author's affiliation

Department of Archaeology
University of Winchester
West Hill, Winchester
Hampshire
SO22 4NR
United Kingdom

# 5. Animal palaeopathology at two Roman sites in southern Britain

Stephanie Vann

## Abstract

*This paper will discuss the results derived from the systematic analysis of palaeopathological lesions from two Roman sites in southern Britain, Alchester and Cups Hotel, Colchester, following the application of a new recording methodology. It will focus particular attention on the cattle (Bos taurus L., 1758) remains from both sites and investigate the potential of animal palaeopathological data to explore questions concerning the economic significance of this species in Roman Britain.*

## Introduction

Previous publications have set out the archaeological rationale behind developing a generic methodology to enable the consistent recognition, recording and description of animal palaeopathological data (O'Connor 2000; Vann 2008; Vann and Thomas 2006). As has been noted by several authors (Albarella 1995, 699; Shaffer and Baker 1997, 256; Thomas and Mainland 2005, 1-2; Vann and Thomas 2006, 1), a number of methodological issues have contributed to the relative neglect of animal palaeopathology as a discipline. However, such issues are not insurmountable and should not prevent the discipline from moving away from an 'interesting' pathological specimen approach of which it has been guilty in the past. As noted by Thomas and Mainland (2005 1-7), such an approach, not only fails to integrate pathological data with the other faunal evidence, but also fails to consider the wider archaeological relevance of palaeopathological lesions. No evidence at a population-level can be gained from merely studying isolated examples (Vann and Thomas 2006, 1).

This study aims to demonstrate the feasibility of population-level analysis through the application of a generic methodology for recording animal palaeo-pathology that has recently been developed (Vann 2008; Vann and Thomas 2006) to two Roman sites in southern Britain. Particular attention will be paid to the cattle (*Bos taurus* L., 1758) remains from both sites in order to investigate the potential of animal palaeopathological data to explore questions concerning their economic significance in Roman Britain.

## Materials

The material selected for this analysis derived from two Roman sites in southern Britain: Alchester, Oxfordshire, and Cups Hotel, Colchester, Essex. These assemblages were not chosen because unusual proportions of palaeopathology were already known to be present. Instead, they were selected because of their availability to the researcher and because they did not appear to be in any way 'extraordinary': they represented the sort of assemblage that any archaeozoologist might be asked to report upon. The fact that both sites are Roman in date also permits inter-site comparison and raises the possibility of novel insights being revealed regarding the economy, status and cultural attributes of the sites.

As one of the earliest Roman bases in inland Britain, with only a single generation of occupation, and given the richness of the organic remains from the site, Thomas (2008) has argued that Alchester has unique potential to illuminate the nature of the diet and provisioning arrangements of the Roman army during the first years of the conquest. The assemblage from Colchester, meanwhile, has the potential to shed light on diet and meat supply in Roman towns. Since these issues are extensive, however, in this case study I will focus specifically upon the types of information that can be gained through a systematic examination of the pathological conditions present within the assemblages.

Questions that will be addressed include:

- What is the relative frequency of palaeo-pathological lesions amongst the different species represented in the assemblages?
- What is the skeletal distribution of these palaeopathological lesions?
- What types of palaeopathological lesions were most prevalent?
- What do the relative frequencies of the palaeopathological lesions reveal about animal health and disease?
- What can the abundance of palaeopathological lesions reveal about the economic significance of cattle at Alchester and Colchester?

## The Alchester assemblage

Alchester, a Roman small town located ten miles north of Oxford and two miles south of Bicester, UK, (Fig. 5.1) has been the subject of investigation since 1996, first under the auspices of Oxford University Archaeological

Society, then through the Universities of Leicester, Oxford and Edinburgh (Sauer 2005, 168). Excavation focussed initially on what was later interpreted as a Roman military parade ground and marching camp near the later Roman town, and then on an annexe of a large military compound, which was shown to be surrounded by a characteristic V-shaped ditch (Sauer and Crutchley 1998, 35). It was common practice for the Roman army, when operating in enemy territory, to build marching camps against surprise attacks, even when spending only one night. However, the comparative frequency of early artefacts from the site suggests that the camp may have existed for a longer time span, perhaps serving as winter quarters (Sauer and Crutchley 1998, 36).

Alchester, which later occupied an important road junction, was in a key strategic position. Situated in the border region of the Catuvellauni and Dobunni tribes, it was in an ideal position to exercise control over wide areas and to obtain sufficient food supplies for the winter (Sauer and Crutchley 1998, 36). It is possible, therefore, that this site represents a so-called 'vexillation fortress', a term coined by Frere and St. Joseph (1974) for forts of 8-12 hectares in size, with an estimated garrison of 2,500-3,000 men (Sauer 2001a, 21). However, the discovery of a tombstone belonging to a legionary soldier raises the possibility that the site was in fact a legionary fortress, and thus housed a much larger cohort of soldiers (Sauer 2005).

Investigation of what was believed to be the front gate, or *Porta Praetoria*, of the fortress revealed two wooden gateposts preserved by waterlogging. Dendrochronology gave both a felling date of between October AD 44 and March AD 45 (Sauer 2001b, 191). However, there is now increasing evidence that this gate belongs to an annexe and is thus later than the fortress, which could date to AD 43 (Sauer 2002, 355) – the year of the successful incorporation of Britain into the Roman Empire. This site, therefore, provides some of the earliest occupational evidence for the Romans in Britain. The date of abandonment of the site is placed before the death of Emperor Nero in AD 68 based upon numismatic evidence (Sauer 2001b, 191). Thus, occupation of the annexe may have lasted for only 25 years. Following the abandonment of the fortress, the town of Alchester began to develop with occupation stretching to the fourth century AD.

Previous analyses of the faunal remains from Alchester (Grant 2001; Powell and Clark 2001; Thomas 2008) have demonstrated that the assemblage is dominated by the principal domestic species – cattle, sheep (*Ovis aries* L., 1758), pig (*Sus scrofa* L., 1758), dog (*Canis familiaris* L., 1758) and horse (*Equus caballus* L., 1758) – although a small number of bones from wild animals, including red deer (*Cervus elaphus* L., 1758), and birds were also recovered (Grant 2001, 63). For this study a sample of 1318 fragments was analysed from the site. Animal bones were recovered from early military phases and later civilian occupation (based on provisional phasing); however, for the purposes of this analysis these have been combined to provide a sufficiently large sample to facilitate inter-site comparison.

## The Cups Hotel assemblage

One of the most significant results of large-scale excavations in Romano-British towns since the 1970s has been the discovery that several of these towns were re-used redundant military fortresses (Crummy 1982, 125). Amongst the towns now known to have had origins closely bound up with the military bases they succeeded is Colchester, whose military defences were levelled so that the new town could cover an area larger than the original fortress (Crummy 1982, 125). As a result, Colchester was largely unprotected during the Boudiccan revolt of AD 60-61, and may consequently have suffered to a greater extent during the uprising (Crummy 1982, 125).

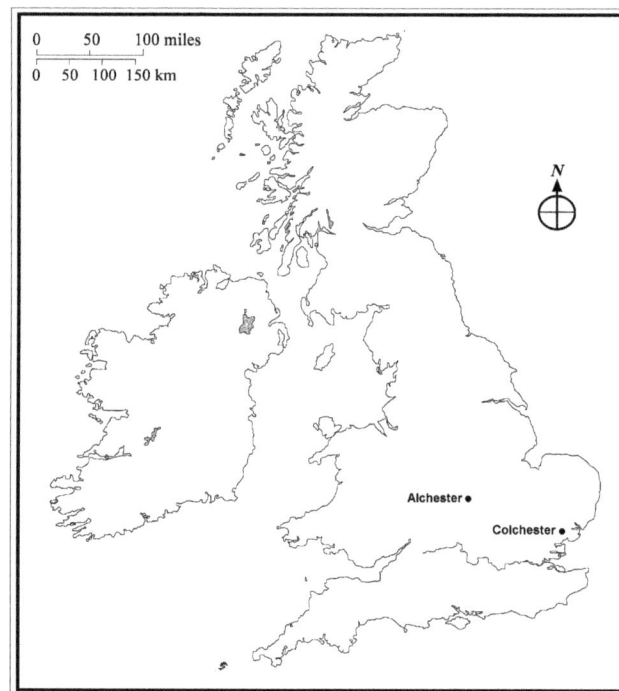

**Figure 5.1**: map showing location of Alchester and Colchester.

Colchester was an important settlement in the Roman and later periods, both because of its status as a port and because of its location in rich farming country (Luff 1993, 7). During the past few decades, intensive excavation by the Colchester Archaeological Trust has yielded significant results from both inside and outside the town walls (Luff 1993, 7). The site of the Cups Hotel was of potential archaeological interest because it lay in the centre of the Roman and medieval town (Crummy 1992, 328). Permission was gained for a rescue excavation in 1973 after the demolition of the hotel and before further construction work began (Crummy 1992, 328). Extensive cellars were discovered to have already removed much of the archaeological remains; however, a north-south strip about six metres wide surviving intact that was the subject of the main phase of excavations (Crummy 1992, 328).

In 1974, after contractors had removed modern cellars on the site, what appeared to be a narrow gravelled road with a cambered surface was revealed in section. The gravel lay on the natural sand and was sealed by typical post-Boudiccan dump-material (Crummy 1977, 84; 1992, 330). It was suggested that this apparently early road indicated the likely presence of the tribune's house close-by, in keeping with equivalent distances in other fortresses at Caerleon and Neuss (Crummy 1977, 84). A section of early Roman wall was also discovered at the site whose timber-framed construction was in-filled with fragments of segmental bricks that showed evidence of burning in AD 60 (Essex Sites and Monuments Record No.: 12298).

The three buildings on the site represented occupation from the early Roman period (*c.* AD 49-60/1) through to the later period (*c.* AD 225-400+; Crummy 1992, 328). A bone comb was found in association with a dispersed hoard of coins and pottery, which securely dated the find to the late Roman rather than Anglo-Saxon period (Crummy 1992, 333). In addition, a cast bronze belt-mount of the type in vogue in the second half of the fourth century AD was discovered in a Norman robber trench.

The previously published faunal assemblage from this site was dominated by the principal domestic species – cattle, sheep, pig, dog and horse – although bones from wild animals, including red deer and roe deer (*Capreolus capreolus* L., 1758), cat (*Felis catus* L., 1758), fish and birds were also recovered. For this study 947 fragments were analysed and all Roman phases were combined.

## Methodology

All material from contexts securely dated to the Roman period from the two sites was subjected to macroscopic examination. Full contextual information, including the data gathered during recent analysis by Thomas (2008) was available for the Alchester assemblage. A similar analysis of the Colchester assemblage was also carried out to ensure that non-pathological data was available in order for prevalence to be calculated. Data gathered for each bone fragment included, where possible: species; skeletal element; side; zone(s) present (Dobney and Reilly 1988); and taphonomic condition (Harland *et al.* 2003). The tooth wear stages of mandibular teeth were recorded using Grant (1982).

During the study of both sites, all potential palaeo-pathological data was inputted into a database following the methodology of Vann (2008), details of which are summarised in Vann and Thomas (2006). Generic terms were used to describe all pathologies. For example, smooth outgrowths of bone were simply noted as 'nodules', a term that includes both enthesophytes and osteophytes, as well as bone outgrowths whose origins were much less clear. This was done with the intention of making the methodology comprehensible to non-specialists, but simultaneously providing sufficient descriptions to cover the great majority of pathologies (Vann and Thomas 2006, 4.3). This also enabled broad categories of disease type to be established and analysed on a regional basis, the value of which has been demonstrated by Murphy (2005). In addition, all

complete cattle metapodia and phalanges, regardless of their stage of deformation, were recorded using the methodology of Bartosiewicz *et al.* (1997). The inclusion of unaltered cattle metapodia and phalanges in the calculation of the pathological index (PI) ensured that the profile generated was a genuine reflection of the frequency of bone deformation at these sites.

## Overview of pathologies

Overall, pathological change was more common at Alchester (11.3%) than Colchester (6.1%). Calculation of the prevalence of palaeopathological lesions by species (Fig. 5.2) reveals that at both sites the vast majority of palaeopathological lesions were found on cattle bones. Dogs, caprines (sheep/goat) and birds were affected with the next greatest frequency, although the dog data for Alchester was biased by the inclusion of partial skeletons; with those remains excluded, the relative proportion of affected dog bones was significantly lower. Palaeo-pathological lesions on pig and horse were only found at Alchester and on red deer at Colchester.

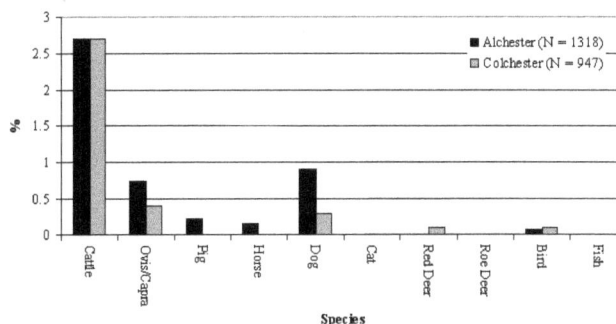

**Figure 5.2**: relative proportions of all palaeopathological lesions by species.

The relative frequencies of major palaeopathological categories by species (Tab 5.1) reveal that the number of cases of palaeopathology was significantly higher at Alchester than at Colchester, both in terms of the total number of cases and the relative frequency.

Inter-species variation is evident within the different palaeopathological categories. Cattle display a greater frequency of bone destruction related pathologies at both sites, whilst the distribution of lesions in caprines is more evenly balanced. However, bone formation related pathologies are more frequent than any other category in dogs and horses. There is also clear inter-site variation. Bone destruction related pathologies are more frequent in pigs at Alchester, whilst instances of bone formation are more frequent in that species at Colchester. The prevalence of palaeopathological lesions on different skeletal elements by species at each site demonstrated both inter-site and inter-species variation.

At Alchester, most major skeletal elements were represented. Of the limb bones, only the femur showed no pathological alteration in any species (Fig. 5.3). This may be a reflection of their relative scarcity – only seven

| Species | Pathology types | Alchester (N=809) | | Colchester (N=947) | |
|---------|-----------------|----------|------|----------|------|
| | | Total Number | % | Total Number | % |
| Cattle | Bone formation | 11 | 3.7 | 0 | 0 |
| | Bone destruction | 15 | 5.1 | 7 | 3.6 |
| | Fracture | 0 | 0 | 0 | 0 |
| Sheep/ goat | Bone formation | 4 | 2.1 | 2 | 0.2 |
| | Bone destruction | 4 | 2.1 | 1 | 0.1 |
| | Fracture | 0 | 0 | 1 | 0.1 |
| Pig | Bone formation | 0 | 0 | 1 | 1.3 |
| | Bone destruction | 3 | 2.5 | 1 | 0.4 |
| | Fracture | 0 | 0 | 0 | 0 |
| Horse | Bone formation | 2 | 1.8 | 0 | 0 |
| | Bone destruction | 0 | 0 | 0 | 0 |
| | Fracture | 0 | 0 | 0 | 0 |
| Dog | Bone formation | 10 | 17.2 | 3 | 2.8 |
| | Bone destruction | 2 | 3.4 | 1 | 2.8 |
| | Fracture | 2 | 3.4 | 0 | 0 |
| Red deer | Bone formation | 0 | 0 | 1 | 5 |
| | Bone destruction | 0 | 0 | 1 | 5 |
| | Fracture | 0 | 0 | 0 | 0 |
| Bird | Bone formation | 0 | 0 | 0 | 0 |
| | Bone destruction | 0 | 0 | 0 | 0 |
| | Fracture | 1 | 5.2 | 0 | 0 |
| | **TOTAL** | **54** | **6.7** | **19** | **0.02** |

**Table.5.1**: numbers and percentages of bones exhibiting different types of palaeopathological alteration by species.

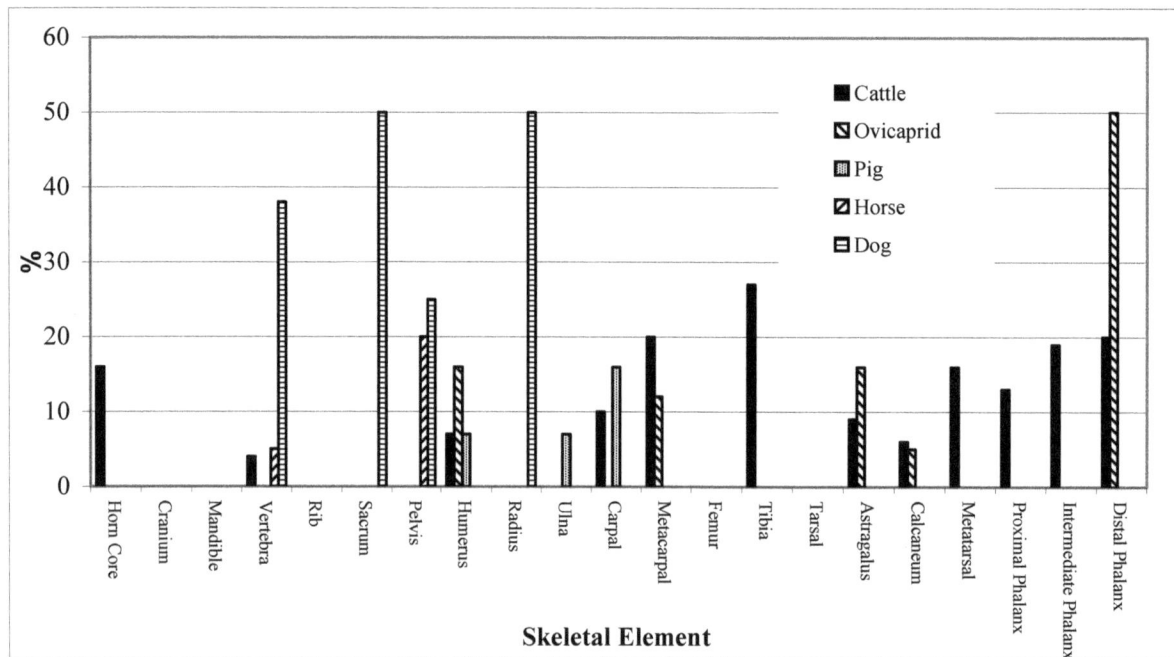

**Figure 5.3**: percentage of skeletal elements affected by palaeopathological lesions at Alchester.

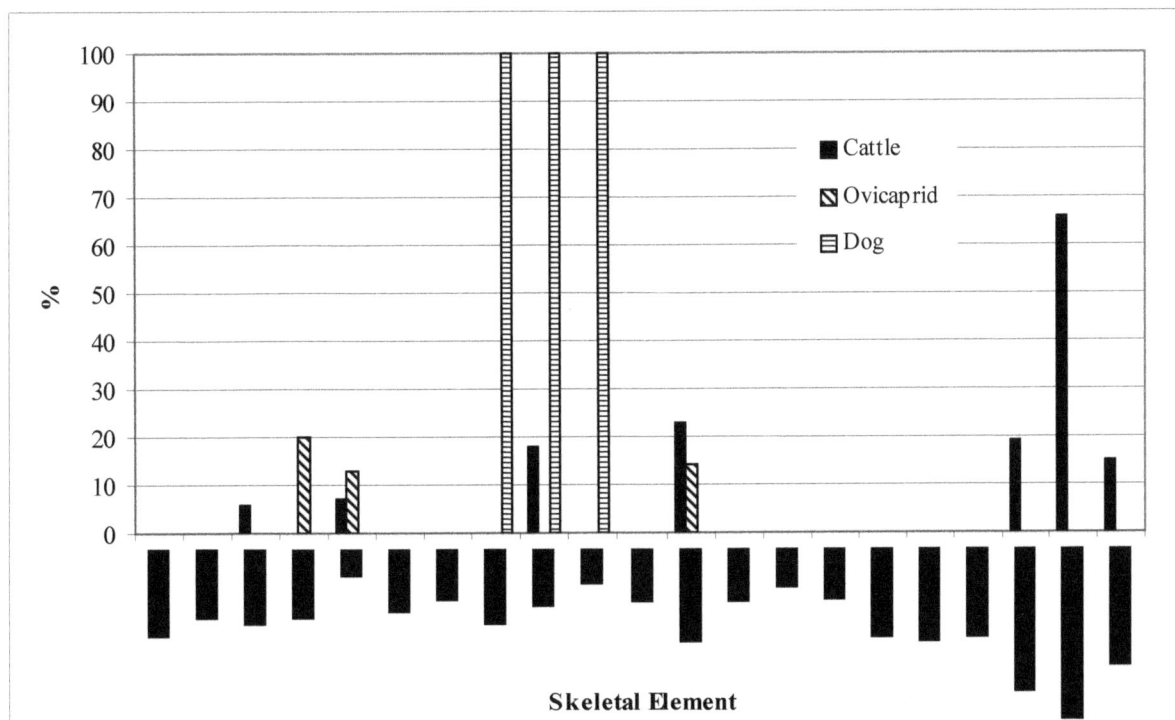

**Figure 5.4**: percentage of skeletal elements affected by palaeopathological lesions at Colchester.

examples out of a total of 293 cattle bones were recovered from the site. In cattle, the vast majority of lesions were found in the feet, particularly the metapodia and phalanges. In caprines, pathological alteration was most frequent in the humerus, which was followed by teeth and then elements of the feet in order of frequency. Pathological alteration in pigs was concentrated in the forelimb elements, whilst horses only showed alteration to the axial skeleton, specifically the pelvis and vertebrae. Lesions were also most frequent in the axial skeleton of dogs although, as previously stated, these results may be skewed due to the presence of articulated material.

At Colchester (Fig. 5.4), only elements of the axial skeleton and forelimbs were affected by palaeo-pathological lesions. The vast majority of those on cattle bones were shown to be present in the fore feet, specifically the metacarpal and phalanges. This is comparable with the results from Alchester where the bones of the feet also showed the most alteration. Palaeopathological lesions in caprines were fewer in number and had less discernible trends, whilst lesions on dog bones were focused entirely on the bones of the forelimb, in particular, the humerus, radius and ulna.

In summary, it can be seen that the prevalence of palaeopathological lesions at both sites was greater in domesticated mammals than in wild species and birds. The number of cases of palaeopathology was significantly higher at Alchester than at Colchester, both in terms of the total number of examples and the relative frequency by species. However, both inter-site and inter-species variation affected the prevalence and skeletal distribution of palaeopathological lesions. Similar conclusions were drawn in a recent review of

palaeopathology at prehistoric and historic sites in Ireland (Murphy 2005). However, at those sites, greater incidences of trauma were noted in both caprines and dogs (Murphy 2005, 21).

## Pathology in cattle

As illustrated in Fig. 5.2, the vast majority of pathological alteration at both sites was found on cattle bones, attention will thus focus on this species.

| | Alchester (N = 293) | | Colchester (N = 195) | |
|---|---|---|---|---|
| | Total Number | % | Total Number | % |
| **Bone formation** | | | | |
| Periostosis | 7 | 2.4 | 0 | 0 |
| Nodule | 4 | 1.4 | 0 | 0 |
| **Bone destruction** | | | | |
| Porosity | 9 | 3.1 | 2 | 1.0 |
| Articular depression or groove | 3 | 1.0 | 5 | 2.6 |
| Cavity | 3 | 1.0 | 0 | 0 |
| **Miscellaneous** | | | | |
| Articular extension | 1 | 0.3 | 0 | 0 |
| Eburnation | 1 | 0.3 | 0 | 0 |
| **TOTAL** | **28** | **9.5** | **7** | **3.6** |

**Table 5.2**: pathological alterations in cattle.

Tab. 5.2 displays the different types of pathology found on cattle bones and their relative frequency within the

31

two assemblages. It can be seen that the number of cases of pathology was significantly higher at Alchester than at Colchester, both in terms of the total number of examples and the relative proportion. Only two types of pathology were recorded at Colchester: porosity and articular depressions or grooves, both bone-destroying lesions. Bone-forming pathologies were entirely absent from cattle bones in this assemblage. In contrast, at Alchester, seven different types of pathology were recorded.

Periostosis, new bone formation following inflammation or infection of the periosteum, was the most common type of bone-forming pathology in the Alchester assemblage. These were all irregularly shaped with indeterminate margins, ranging from less than 100 mm$^2$ up to 3696 mm$^2$ in area. There was no obvious pattern to their distribution: prevalence rates were one out of 14 humeri, two out of 11 tibiae (Fig. 5.5), one out of 370 large-mammal-sized rib fragments, two out of 43 vertebrae and one out of 11 astragali. It is probable that all represent instances of infection or inflammation. It is very difficult to determine the specific causes of such lesions since they can occur as a result of either direct or localised inflammation or infection, consequent to localised or systemic disease processes or trauma. The presence of such lesions on the visceral surface of a rib is interesting because this implies the presence of some form of pleural infection.

The second type of pathological bone formation in the Alchester assemblage was nodular bone growth. These were generally smooth protrusions of bone with a poorly defined margin, and of irregular or round/oval shape. Most lesions were of a small size, ranging from *c.* 6mm$^2$ to 250mm$^2$ in area. All of these were osteophytes found around the proximal articular margins of both anterior and posterior phalanges. Such lesions occur in response to a need for increased joint stability. This can be precipitated by degenerative change to the joint consequent to age, trauma or nutritional deficiency (Baker and Brothwell 1980, 114-117; Peterson 1988). They had prevalence rates of one out of 38 proximal phalanges, two out of 31 intermediate phalanges, and one out of 20 distal phalanges.

Increased porosity was the most common type of bone destruction-related pathology in the Alchester assemblage, and the second most common in the Colchester assemblage. At both sites these lesions were typically irregular in shape with poorly defined margins, ranging from less than 100 mm$^2$ up to 3696 mm$^2$ in area. These were distributed throughout the skeleton, examples being found on two out of ten cranial elements, one out of eleven tibiae, one out of twelve metatarsals, one out of six tarsals, and one out of thirty-one intermediate phalanges at Alchester, as well as one out of eleven radii, and one out of twenty-one metacarpals at Colchester. Porosity was also found in association with other pathological manifestations such as periostosis, nodular bone formation and eburnation at Alchester. At both sites, all of these lesions were of unknown aetiology, although those in association with other conditions may represent additional symptoms of inflammation, infection or degenerative joint disease.

Within the Alchester assemblage, all articular depressions were found on cattle bones: one on the proximal articulation of a lunate and two on proximal phalanges (one each on the proximal and distal articulations).

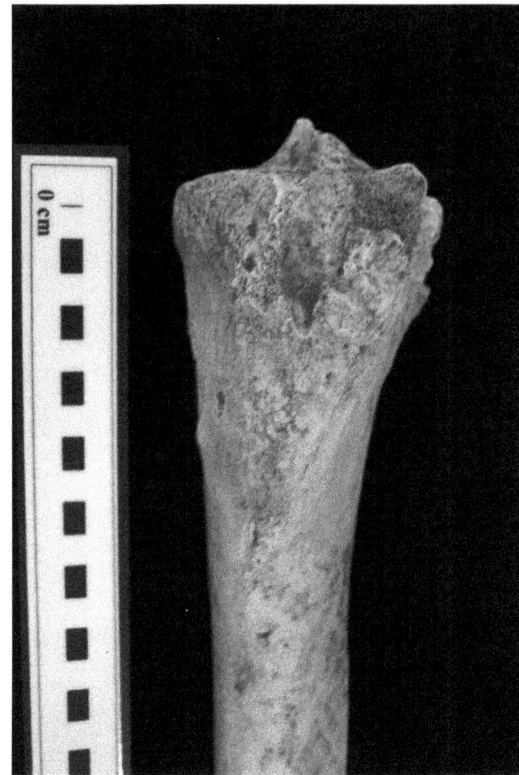

**Figure 5.5**: periostosis on the proximal tuberosity of a cattle tibia from Alchester; anterior view.

These had well-defined margins, and ranged in area from 4-44 mm$^2$. Of the three types of depressions commonly seen on the articular surfaces of cattle phalanges described by Baker and Brothwell (1980, 110), those found in the Alchester assemblage most resemble those classified as types two and three. Type two is "of variable length between the articular facets and are more common on the lower extremity of the second [intermediate] phalanx than at other sites" (Baker and Brothwell 1980, 110). Type three "take the form of a slit of variable length and not inconsiderable depth running across the articular facets in a line slightly oblique to the medio-lateral axis," and are typically found on the distal phalanx (Baker and Brothwell 1980, 110). The condition was found on one out of six cattle tarsals and two out of thirty-eight cattle proximal phalanges.

Articular grooves and depressions were also seen in the Colchester assemblage, representing the most prevalent form of pathological response at that site. However, unlike the Alchester assemblage, these were not found on phalanges or tarsals. Instead, they were found on the proximal articular surfaces of radii and metacarpals with a prevalence of one out of 11 radii and four out of 21 metacarpals in the assemblage. With the exception of their location, these would also appear to conform to types two and three as previously defined by Baker and Brothwell (1980, 110).

Cavities were defined as a hollow area within a bone. These displayed no real pattern, being linear, round/oval or irregular in shape, with examples with both poorly and well-defined margins. They ranged in area from 4 mm$^2$ up to 60 mm$^2$, and were found solely on metapodia and tarsals at Alchester with a prevalence of one out of 10 metacarpals, one out of 12 metatarsals, and one out of six tarsals. Two of these specimens, one on the proximal articulation of a scapho-cuboid (navicular) and one on the distal condyle of a metatarsal, had smooth interior surfaces, suggesting they might be the result of cysts, abnormal sacs in the body, filled with a fluid or semi-solid and enclosed in a membrane. The other, located on the proximal articular surface of a metacarpal, had an irregular interior surface. This may indicate that it was due to an abscess, a collection of pus or other matter contained in a localised area of the body.

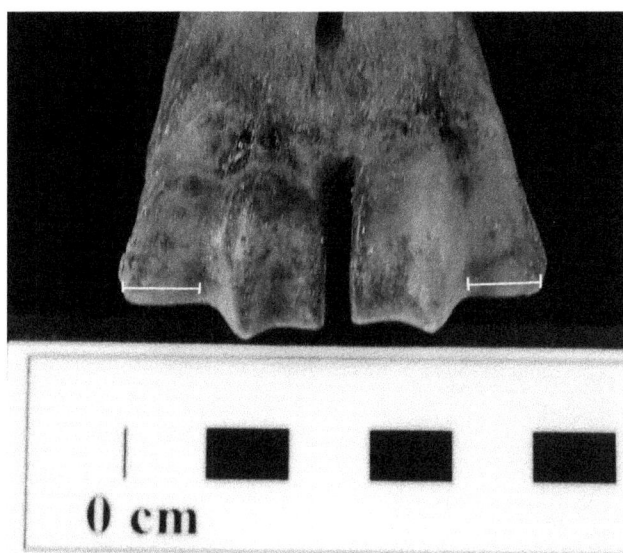

Figure 5.6: slight broadening of the lateral condyle of a cattle metacarpal from Alchester; anterior view.

Eburnation, the degeneration of articular surfaces into a hard, polished, ivory-like surface, was found on a single astragalus at Alchester and could be osteoarthritis, although without the presence of osteophytes or grooving it is impossible to be certain. It does, at the very least, indicate degenerative joint disease and cartilage destruction. The broadening of the distal articulation of a cattle metacarpal (stage two after Bartosiewicz et al. 1997) from Alchester (Fig. 5.6) most likely represents a skeletal adaptation to load-bearing (Bartosiewicz et al. 1993). The prevalence of this condition was one out of a total of 10 cattle metacarpals identified within the assemblage.

## Pathological index

The pathological index devised by Bartosiewicz et al. (1997) was calculated for all complete cattle metapodia and phalanges. No differentiation was made between those that were obviously pathological and those that were not. The aim was to investigate whether or not the cattle at these sites showed any evidence of pathological alteration that might have resulted from their use.

As can be seen from the results summarised in Tab 5.3, the majority (79%) of cattle bones from Alchester showed no pathological alteration, having a pathological index (PI) value of 0.000. Few showed significant deformation, with only one bone, a distal phalanx, displaying a PI of more than 0.200.

These results are comparable to those from the Colchester assemblage (Tab. 5.4). At that site, the majority of autopodia (64%) showed no pathological alteration. Few showed significant deformation, with only one bone, an intermediate phalanx, displaying a PI of more than 0.200.

The average PI value was also calculated for each skeletal element and these results are displayed in Fig. 5.7. It can be seen that the distal phalanx exhibited the highest average PI value in the Alchester assemblage, whilst in the Colchester assemblage the average PI value for the intermediate phalanx was significantly higher than for any other element. Neither site showed any pathological alteration in the metapodia using this methodology.

## Discussion

Whilst it is true that the systematic analysis of palaeopathological data has resulted in the identification of isolated instances of certain pathologies that have limited analytical potential, a number of discernible trends are apparent. In particular, the palaeopathological evidence focuses attention upon the economic significance of cattle at Alchester and Colchester and inter-site variation in herd management.

The prevalence of periostosis is significantly higher in cattle at Alchester than at Colchester. This would suggest that the prevalence of infection or inflammation was also higher at the former site, something that may relate to variation in herd management or environmental conditions. Increased susceptibility to disease can be one consequence of poor nutrition, as can increased intervals between offspring and an increase in the time taken to reach maturity (Davies 2005, 85). It is therefore possible that the greater prevalence of periostosis at Alchester may reflect a higher rate of nutritional deficiency. Such under-nutrition may reflect the degree of access to grazing or surplus feed, as well as the presence of parasites in the alimentary canal, which can reduce the efficiency with which food is digested and interferes with the absorption of the products of digestion (O'Connor 2000, 102).

Other potential causes of increased rates of infection and inflammation include trauma and environmental conditions. An examination of the evidence from historic period sites in Ireland, for example, demonstrated that infection was common on lower leg bones, suggesting that these elements were most susceptible to infectious diseases such as foot root (Murphy 2005, 16).

Such infections have been linked by other authors to prolonged stalling or keeping animals on soft and muddy

| PI | Metacarpal | Metatarsal | Proximal Phalanx | Intermediate Phalanx | Distal Phalanx | Total |
|---|---|---|---|---|---|---|
| 0 | 3 | 1 | 20 | 14 | 10 | **48** |
| 0.091 | 0 | 0 | 3 | 4 | 0 | **7** |
| 0.143 | 0 | 0 | 0 | 0 | 3 | **3** |
| 0.181 | 0 | 0 | 0 | 2 | 0 | **2** |
| 0.272 | 0 | 0 | 0 | 0 | 0 | **0** |
| 0.285 | 0 | 0 | 0 | 0 | 1 | **1** |
| **TOTAL** | **3** | **1** | **23** | **20** | **14** | **61** |

**Table 5.3**: summary of the pathological index data from Alchester, based on methodology devised by Bartosiewicz *et al.* (1997).

| PI | Metacarpal | Metatarsal | Proximal Phalanx | Intermediate Phalanx | Distal Phalanx | Total |
|---|---|---|---|---|---|---|
| 0 | 0 | 0 | 19 | 5 | 8 | **32** |
| 0.091 | 0 | 0 | 4 | 6 | 0 | **10** |
| 0.143 | 0 | 0 | 0 | 0 | 3 | **3** |
| 0.181 | 0 | 0 | 1 | 3 | 0 | **4** |
| 0.272 | 0 | 0 | 0 | 1 | 0 | **1** |
| 0.285 | 0 | 0 | 0 | 0 | 0 | **0** |
| **TOTAL** | **0** | **0** | **24** | **15** | **11** | **50** |

**Table 5.4**: summary of the pathological index data from Colchester, based on methodology devised by Bartosiewicz *et al.* (1997).

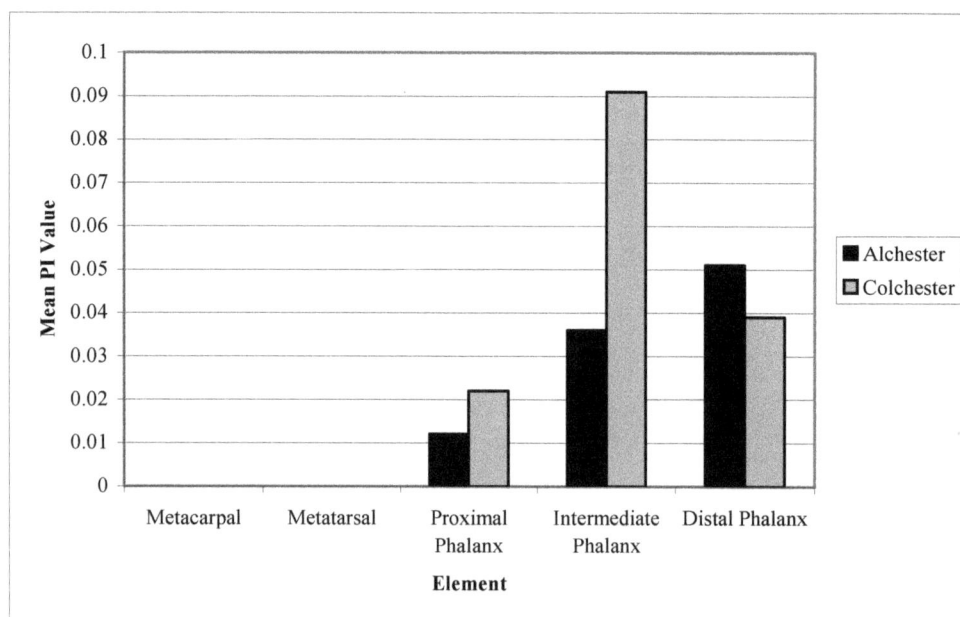

**Figure 5.7**: comparison of mean pathological index values by skeletal element at Alchester and Colchester.

pastures (Baker and Brothwell 1980: 73) and this could explain the periosteal lesions found on some lower limb bones found at Alchester and Colchester.

At both sites the majority of pathological alterations affected the lower limb bones, particularly the metapodia and phalanges. This may in part be a function of taphonomy since the elements associated with locomotion undergo more stresses than other elements and are thus composed of denser bone (Shaffer and Baker 1997, 259). However, the nature of the pathological alteration was not the same at both sites. Osteophytosis and eburnation for example, both often associated with degenerative joint disease, were noted at Alchester, but not Colchester. Deformation of these types has been linked to the use of cattle for traction (Bartosiewicz *et al.* 1997; de Cupere *et al.* 2000; Fabiš 2005; Higham *et al.* 1981; Johannsen 2005). This variation may thus reflect differences between the two sites – one mainly military and one mainly urban – and the nature of the supply of animals to them.

The infrequency of osteophytosis and eburnation seen at Alchester is supported by the results from those bones recorded using the methodology devised by Bartosiewicz *et al.* (1997). When compared with the results of other studies, is it evident that the mean PI values for the Alchester assemblage (0.028) and for the Colchester assemblage (0.046) are significantly lower than the data published by de Cupere *et al.* (2000, 261).

One factor that could have determined the severity of deformation in the lower limbs was the landscape that the cattle inhabited. The physiography of the region around Alchester is dominated by clay lowland, lying mostly at less than 80m above OD (Ordnance Datum). This is divided into two north-eastward-trending vales separated by a range of hills capped by the limestones and sands of the Corallian, Portland, Purbeck and Whitchurch Sand formations, which cross diagonally from near Oxford to Aylesbury (British Geological Survey 1995, 1-2). Further to the east, around Colchester, Quaternary deposits are more commonplace, particularly Pre-Anglian fluvial deposits (British Geological Survey 1996, Fig. 29), whilst the central part of the London Basin is in-filled with Palaeogene deposits, dominated by sand and mudstone, the London Clay Formation being the thickest and most widespread of these (British Geological Survey 1996, 3-4).

Even today, this region of the Thames Valley is largely agricultural (British Geological Survey 1995, Preface), and it does not seem unreasonable to suppose that these lowlands were also exploited in this manner during earlier periods. If so, such terrain would presumably put less stress upon the limbs of an individual animal than would be seen at the other study sites, such as Sagalassos on the rocky hill slope of the western Taurus range, or Liberchies, Namur and Torgny, where the topography was also generally hilly (de Cupere *et al.* 2000, 258-260). It is possible, therefore, that this may have contributed in some way to the observed differences at Alchester and Colchester.

Nutritional deficiency has been implicated as a factor in other large mammals such as moose (Peterson 1988). The 'early nutrition hypothesis' suggests that subtle developmental abnormalities in cartilage due to under-nutrition subsequently results in a higher frequency of degenerative lesions in later life (Peterson 1988, 465). As discussed previously, poor nutrition can also result in a greater susceptibility to infection, something seen at Alchester but not Colchester. However, the PI evidence that reveals a higher value for Colchester contradicts this hypothesis.

Age can also be an important factor: degenerative change occurs less frequently in younger individuals (Armour-Chelu and Clutton-Brock 1985, 300). The collections from Roman Turkey and Belgium (de Cupere *et al.* 2000, 264) differed from the modern Romanian assemblage used by Bartosiewicz *et al.* (1997) in that they contained both male and female individuals. This increased the likelihood that the sample contained animals that were never put to work. The same may also be true of the assemblages from Alchester and Colchester, and may have contributed to the low PI values.

In order to explore the link between lower limb pathology in cattle and age, the mortality profiles based upon tooth eruption and wear were examined for the two sites. The stages were based upon those defined by O'Connor (2003, Table 31). The total number of mandibles recorded from both Alchester and Colchester was too low to draw any definitive conclusions. Nonetheless, there was a greater frequency of adult or elderly cattle mandibles at Alchester than there were mandibles belonging to younger individuals. At Colchester, no adult cattle were recorded; all were juvenile.

Since the low numbers of recordable mandibles had almost certainly biased the results, mortality profiles were constructed using post-cranial fusion data (Fig. 5.8). These data demonstrate that only 11% of the total number of cattle at Alchester was still unfused by the late-fusing stage, whilst at Colchester only 2.5% of cattle were still unfused at this stage. This would indicate that the majority of animals at both sites were above 42 months of age (Reitz and Wing 1999, Table 3.5) when they were slaughtered, although a greater number of mature animals were present within the Colchester assemblage.

The preponderance of older animals is worthy of attention, in light of the low rate of pathology present among the cattle at both sites and given that degenerative joint disease is linked to senescence. However, post-cranial fusion data does not have the resolution necessary to determine whether the cattle were slaughtered soon after they had reached their maximum size or kept for many years beyond.

The infrequency of pathological change in the bones of the lower limb may instead reflect the lifestyle of the individual animals present within the Alchester and Colchester populations. The use of cattle for traction, which has been shown to exacerbate degenerative joint disease, may have been low at both these sites. This information can be used to enhance our understanding of the way in which animals were supplied to these sites. Other authors have noted that the highest concentrations of adult cattle in Romano-British samples have appeared on military and urban settlements (Maltby 1981, 182). Organisation of cattle marketing and the need to provision these centres with meat may have resulted in the supply of particular types and age groups of cattle (Maltby 1981, 182).

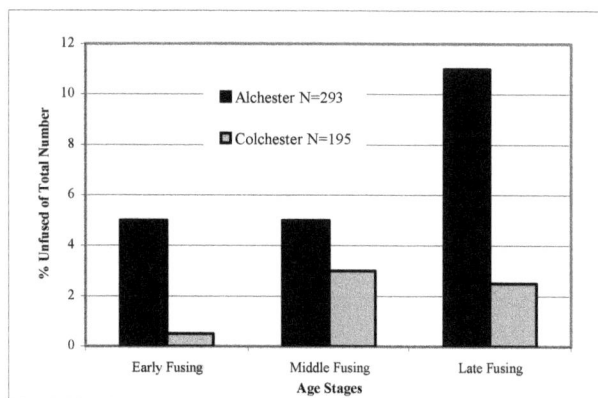

**Figure 5.8**: frequency of unfused cattle post-cranial bones (stages after Reitz and Wing 1999, Table 3.5).

The early date of the fortress at Alchester means that examination of the nature of meat supply has the potential to shed light on the nature of relations with indigenous communities. Unable to supply its own meat, the garrison would have had to rely on local supply, but what was the nature of that supply? Did the Romans forcibly requisition, was it taxed, or was a system of market exchange in operation at the very earliest conquest period (Thomas 2008)?

Evidence from other sites suggests that the earliest Roman settlements were established rapidly following the Claudian invasion in AD 43, and one of the first tasks of the invading force must have been to ensure adequate food supplies (Grant 2004, 372). According to Tacitus, every Roman fort in Britain, when Agricola was governor in AD 78-84, was provided with sufficient sup-plies to last for a year (Davies 1971, 122-123). Calculations show that each Roman soldier would eat approximately one-third of a ton of corn each year (Davies 1971, 123). The establishment of a new, demanding and powerful group of consumers within local farming communities therefore had the potential to bring about considerable disruption (Grant 2004, 372). However, if the army intended to remain for any length of time, they would need to ensure that local animal husbandry and agriculture were not fatally undermined (Grant 2004, 372).

In peacetime, the army used various sources from which to obtain its food supplies. One major source was the civilians of the province; this could take the form of requisitions or compulsory purchase at a fixed price (Davies 1971, 123). Another important source was food produced on military land (*territorium* or *prata*), which extended for a considerable distance around each fort; this was sometimes grown by the military themselves, at other times by civilians to whom the land was leased (Davies 1971, 123). Contracts for supplies in bulk were also used (Davies 1971, 123).

Assemblages from early fortresses included higher proportions of cattle than were common at contemporary native settlements (Grant 2004, 372; King 1999, 179-180), but the beef consumed was mainly from old or even elderly animals (Grant 2004, 372). Sites such as the Flavio-Trajanic auxiliary fort of Leucarum at Loughor in West Wales have shown that cereals and vegetable products provided the basis of the army diet; however, cereal cultivation required cattle for ploughing and other agricultural tasks (Grant 2004, 372). Thus, full-grown cattle may have been preferentially chosen for slaughter, both to ensure herds remained viable and to permit animals to be trained for the plough (Grant 2004, 372).

The mortality profile of the cattle from Alchester is consistent with the pattern established at other early fortress sites. However, the low PI values and minimal evidence for pathological deformation of skeletal elements previously associated with traction challenges the assumption that such activity was a major role for local cattle herds and the reason for their mature age of slaughter. Whilst the evidence does not entirely rule out their use for traction in this area, such activity, if it occurred, would have had to have been sufficiently infrequent that it caused little or no associated pathological change. It would seem more likely that the mortality profile reflects the demand for hides to be manufactured into leather, something that the military required in significant quantities (Grant 2004, 372) and which others (*e.g.* Applebaum 1958, 74-75) have suggested was, along with beef, the main economic use for cattle during this period. Alternatively, the Roman military may have been selective, demanding the supply of cattle not crippled by activity such as traction from the local population.

Cattle were also the most important species in respect to their contribution to the Romano-British diet at urban centres (Grant 2004, 377). Deposits from towns such as Chichester, Winchester, Silchester, Leicester and Lincoln point to the development of centralised processing and distribution (Grant 2004, 377). The growth of large urban centres made the rearing of cattle for beef an increasingly viable strategy (Grant 2004, 377). Concentrated populations of people could share the meat and offal provided by each of these large animals. However, large-scale exploitation required an efficient system for carcass processing, meat distribution and preservation (Grant 2004, 377). Networks that had been established by the military as a means of provisioning early settlements, such as that at Alchester, would doubtless have been subsequently exploited to supply growing urban populations (Grant 2004, 377). Evidence for this can be seen in the mortality profiles of many urban deposits that exhibit narrow age ranges which reflect a continuation of the culling practices used by the military (Grant 2004, 377).

An increased investment in cattle could also reflect an expansion of cereal cultivation and the increased use of cattle as draught animals (Grant 2004, 377). Dumps of primary butchery waste from urban sites such as Dorchester, Cirencester and Lincoln demonstrate that the majority of the cattle being slaughtered were mature, *i.e.* older than four years, suggesting that they may have had a working life prior to being eaten (Cool 2006, 85). This would be comparable with the data from the extramural settlement at Alchester (Powell and Clark 2001, 401-402), which observed a high proportion of old animals in the assemblage. In addition, pathological alteration of cattle phalanges, including articular extension and lesions in the articular surface, were linked to sustained high levels of stress upon the animals (Powell and Clark 2001, 413-414). However, whilst the PI value for Colchester was higher than that of Alchester, indicating greater pathological alteration at the solely urban site, the value was still very low in comparison to previously published comparative data (de Cupere *et al.* 2000, 261). This may be due to inter-site variation and reflect aforementioned differences in topography or age, Sagalassos in particular is noted as having an "old slaughter age" (de Cupere *et al.* 2000, 259). Nonetheless, it was not possible to see the increased use of cattle for traction in the data set.

The data from Cups Hotel was comparable to other Colchester assemblages of the Roman period. At those sites the cattle were also mainly mature. However, approximately 1% exhibited lesions along the medial edge of the acetabulum, which were considered to stem from over-rotation of the hip during traction (Luff and Brothwell 1993, 105). This was not seen on the cattle at Cups Hotel. This suggests, therefore, that at least some cattle were used for traction in the Colchester area, even if these animals were not found at the Cups Hotel site. Such inter-site variation may reflect variations in supply and demand in different parts of the town.

## Conclusions

The case studies illustrated in this paper demonstrate the usefulness of the systematic analysis of palaeopathological data (Vann 2008). Both of these assemblages were 'typical' Romano-British assemblages; the prevalence of pathologies was generally low. Nonetheless, their study has provided fresh insights into the human-animal relationships at two sites of interest to Roman archaeology. The calculation of the pathological index alone produced some interesting conclusions that contribute to our understanding of meat supply to these sites. There is little evidence for the use of cattle as draught animals at either site, suggesting that they were primarily intended for consumption or leather manufacture.

The systematic analysis of many forms of pathology has also revealed clear differences in the prevalence of particular types of pathology, *e.g.* inflammation and infection, although it is not possible to conclusively establish the causes of these conditions. Hopefully future applications of a systematic approach will provide further population-based studies with which

the data from Alchester and Colchester can be compared. Such comparisons would not only provide further insight into animal husbandry in Roman Britain and elsewhere, but would also enable the significance of particular lesions to be assessed and thus stimulate more detailed research into those conditions which reveal significant patterning.

## Acknowledgements

I would like to thank Dr. Eberhard Sauer of the School of History, Classics and Archaeology, University of Edinburgh and Dr. Richard Thomas and Tony Gouldwell of the School of Archaeology and Ancient History, University of Leicester, for permitting me to have access to the Alchester and Cups Hotel, Colchester, assemblages. I am also grateful to Dr. Richard Thomas for his comments on an earlier draft of this text.

## Bibliography

Albarella, U. 1995. Depressions on sheep horncores. *Journal of Archaeological Science* 22, 699-704.

Applebaum, S. 1958. Agriculture in Roman Britain. *Agricultural History Review* 6 (2), 66-86.

Armour-Chelu, M. and Clutton-Brock, J. 1985. Notes on the evidence for the use of cattle as draught animals at Etton, pp. 297-302, in Pryor, F., French, C. and Taylor, M. (eds), An interim report on excavations at Etton, Maxey, Cambridgeshire, 1982-1984. *The Antiquaries Journal* 65, 275-311.

Baker, J. and Brothwell, D. 1980. *Animal Diseases in Archaeology*. London: Academic Press.

Bartosiewicz, L., Van Neer, W. and Lentacker, A. 1993. Metapodial asymmetry in draft cattle. *International Journal of Osteoarchaeology* 3, 69-75.

Bartosiewicz, L., Van Neer, W. and Lentacker, A. 1997. *Draught Cattle: their Osteological Identification and History*. Annales Sciences Zoologiques Vol. 281. Tervuren: Musée Royal de L'Afrique Centrale.

British Geological Survey. 1995. *Geology of the Country around Thame: Memoir for 1:50,000 Geological Sheet 237 (England and Wales)*. London: HMSO.

British Geological Survey. 1996. *British Regional Geology: London and the Thames Valley* (fourth edition). London: HMSO.

Cool, H. E. M. 2006. *Eating and Drinking in Roman Britain*. Cambridge: Cambridge University Press.

Crummy, P. 1977. Colchester: The Roman fortress and the development of the colonia. *Britannia* 8, 65-105.

Crummy, P. 1982. The origins of some major Romano-British towns. *Britannia* 13, 125-134.

Crummy, P. 1992. *Excavations at Culver Street, the Gilberd School, and Other Sites in Colchester*

*1971-85.* Colchester Archaeological Report 6. Colchester: Colchester Archaeological Trust.

Davies, J. J. 2005. Oral pathology, nutritional deficiencies and mineral depletion in domesticates – a literature review, pp. 80-88, in Davies, J., Fabiš, M., Mainland, I., Richards, M. and Thomas, R. (eds). *Diet and Health in Past Animal Populations: Current Research and Future Directions.* Oxford: Oxbow Books.

Davies, R. W. 1971. The Roman military diet. *Britannia* 2, 122-142.

de Cupere, B., Lentacker, A., Van Neer, W., Waelkens, M. and Verslype, L. 2000. Osteological evidence for the draught exploitation of cattle: first applications of a new methodology. *International Journal of Osteoarchaeology* 10, 254-267.

Dobney, K and Reilly, K. 1988. A method for recording archaeological animal bones: the use of diagnostic zones. *Circaea* 5 (2), 79-96.

Fabiš, M. 2005. Pathological alteration of cattle skeletons – evidence for the draught exploitation of ani-mals?, pp. 58-62, in Davies, J., Fabiš, M., Mainland, I., Richards, M. and Thomas, R. (eds), *Diet and Health in Past Animal Populations: Current Research and Future Directions.* Oxford: Oxbow Books.

Frere, S. S. and St. Joseph, K. 1974. The Roman fortress at Longthorpe. *Britannia* 5, 1-129.

Grant, A. 1982. The use of tooth wear as a guide to the ageing and sexing of domestic animals, pp. 91-108, in Wilson, B., Grigson, C. and Payne, S. (eds), *Ageing and Sexing Animal Bones from Archaeological Sites.* Oxford: British Archaeological Reports British Series 109.

Grant, A. 2001. Preliminary note on the animal bone remains from Alchester, pp. 63-64, in Sauer, E. (ed.), Alchester, a Claudian 'Vexillation Fortress' near the western boundary of the Catuvellauni: new light on the Roman invasion of Britain. *The Archaeological Journal* 157, 1-78.

Grant, A. 2004. Animals and the economy and ideology of Roman Britain, pp. 371-392, in Todd, M. (ed.), *A Companion to Roman Britain.* Oxford: Blackwell Publishing.

Harland, J. F., Barrett, J. H., Carrott, J., Dobney, K. and Jaques, D. 2003. The York System: an integrated zooarchaeological database for research and teaching. *Internet Archaeology* 13.

Higham, C. F. W., Kijngam, A., Manly, B. F. J. and Moore, S. J. E. 1981. The bovid third phalanx and prehistoric ploughing. *Journal of Archaeological Science* 8, 353-65.

Johannsen, N. 2005. Palaeopathology and Neolithic cattle traction: methodological issues and archaeological perspectives, pp. 39-51, in Davies, J., Fabiš, M., Mainland, I., Richards, M. and Thomas, R. (eds), *Diet and Health in Past Animal Populations: Current Research and Future Directions.* Oxford: Oxbow Books.

King, A. 1999. Diet in the Roman world: a regional inter-site comparison of the mammal bones. *Journal of Roman Archaeology* 12, 168-202.

Luff, R. 1993. *Animal Bones from Excavations in Colchester, 1971-85.* Colchester Archaeological Report 12. Colchester: Colchester Archaeological Trust and English Heritage.

Luff, R and Brothwell, D. 1993. Health and Welfare, pp. 101-126, in Luff, R. (ed.), *Animal Bones from Excavations in Colchester, 1971-85.* Colchester Archaeological Report 12. Colchester: Colchester Archaeological Trust and English Heritage.

Maltby, M. 1981. Iron Age, Romano-British and Anglo-Saxon animal husbandry – a review of the faunal evidence, pp. 155-203, in Jones, M. and Dimbleby, G. (eds), *The Environment of Man: the Iron Age to the Anglo-Saxon Period.* Oxford: British Archaeological Reports British Series 87.

Murphy, E. 2005. Animal palaeopathology in prehistoric and historic Ireland: a review of the evidence, pp. 8-23, in Davies, J., Fabiš, M., Mainland, I., Richards, M. and Thomas, R. (eds), *Diet and Health in Past Animal Populations: Current Research and Future Directions.* Oxford: Oxbow Books.

O'Connor, T. 2000. *The Archaeology of Animal Bones.* Stroud: Sutton Publishing.

O'Connor, T. 2003. *The Analysis of Urban Animal Bone Assemblages: a Handbook for Archae-ologists.* The Archaeology of York 19/2. York: York Archaeological Trust and the Council for British Archaeology.

Peterson, R. O. 1988. Increased osteoarthritis in moose from Isle Royale. *Journal of Wildlife Disease* 24 (3), 461-466.

Powell, A. and Clark, K. M. 2001. Animal Bone, pp 395-416, in Booth, P. M., Evans, J. and Hiller, J. (eds), *Excavations in the Extramural Settlement of Roman Alchester, Oxfordshire, 1991.* Oxford Archaeology Monograph No. 1. Oxford: Oxford Archaeological Unit.

Reitz, E. J, and Wing, E. S. 1999. *Zooarchaeology.* Cambridge: Cambridge University Press.

Sauer, E. 2001a. Alchester, a Claudian 'Vexillation Fortress' near the western boundary of the Catuvellauni: new light on the Roman invasion of Britain. *The Archaeological Journal* 157, 1-78.

Sauer, E. 2001b. Alchester Roman fortress. *Current Archaeology* 173, 189-191.

Sauer, E. 2002. The Roman invasion of Britain (AD 43) in imperial perspective: a response to Frere and Fulford. *Oxford Journal of Archaeology* 21 (4), 333-363.

Sauer, E. 2005. Alchester: in search of Vespasian. *Current Archaeology* 196, 168-176.

Sauer, E. and Crutchley, S. 1998. Alchester: a Roman fort and parade ground? *Current Archaeology* 157, 34-37.

Shaffer, B. S, and Baker, B. W. 1997. Historic and pre-historic animal pathologies from North America. *Anthropozoologica* 25, 255-261.

Thomas, R. 2008. Supply-chain networks and the Roman invasion of Britain: a case study from Alchester, Oxfordshire, pp. 31-51, in Stallibrass, S. and Thomas, R. (eds), *Feeding the Roman Army: the Archaeology of Production and Supply in NW Europe*. Oxford: Oxbow.

Thomas, R. and Mainland, I. 2005. Introduction: animal diet and health – current perspectives and future directions, pp. 1-7, in Davies, J., Fabiš, M., Mainland, I., Richards, M. and Thomas, R. (eds), *Diet and Health in Past Animal Populations: Current Research and Future Directions*. Oxford: Oxbow.

Vann, S. 2008. *Recording the Facts: a Generic Recording System for Animal Palaeopathology*. Unpublished PhD thesis. Leicester: University of Leicester.

Vann, S. and Thomas, R. 2006. Humans, other animals and disease: a comparative approach towards the development of a standardised recording protocol for animal palaeopathology. *Internet Archaeology* 20 (http://intarch.ac.uk/journal /issue20/vannthomas_index.html).

## Author's affiliation

School of Archaeology and Ancient History
University of Leicester
University Road
Leicester LE1 7RH
United Kingdom

# 6. Understanding past human-animal relationships through the analysis of fractures: a case study from a Roman site in The Netherlands

Maaike Groot

## Abstract

*In studying fractures in archaeology, we should focus on what they can tell us about human-animal relationships. It is important to show other (zoo-) archaeologists that palaeopathology can be a valuable tool in answering (zoo-) archaeological questions. In this paper, a short summary of fracture types, healing, and complications is given and the problems and possibilities of studying fractures in palaeopathology are discussed. Nineteen fractures from a Roman-period site in The Netherlands are then presented. Fracture prevalence rates for this site are discussed and compared with currently published data. Possible explanations for the high fracture rate in dogs are discussed, including maltreatment by humans and work-related injuries.*

## Introduction

Fractures are one of the most common types of pathology seen in archaeological assemblages of animal bones (Ortner 2003, 119); they are also one of the easiest to recognise and diagnose. Unless they are very well-healed, fractures will be recognisable to all but the most inexperienced zooarchaeologist. This makes it less likely that this type of pathology will be overlooked, and this means that, taphonomic factors aside, the number of fractures for different samples can be regarded as broadly representative of the prevalence of fractures within a living population; although the numbers of well-healed green-stick and stress fractures are likely to be underestimated (see below).

Fractures in animal bones have not received the attention they deserve and there are few publications that deal specifically with their occurrence (Gal this volume; Teegen 2005; Udrescu and Van Neer 2005). Because of the scarcity of publications, this study largely makes use of palaeopathological literature on human fractures. However, while the biological responses to fractures, such as the healing process and subsequent complications, will be similar for all mammals, it is recognised that their frequency and distribution is not comparable because of the differences in anatomical position and activities between humans and other mammals.

This paper will detail the animal bone fractures recorded from an archaeological site in The Netherlands: Tiel-Passewaaij (Fig. 6.1). The site dates to the Roman period and is situated within the borders of the Roman Empire. This paper will be concerned with the animal bones from two rural settlements in this location. No fractures were found in wild animals; therefore only domestic mammals are discussed. The total number of identified domestic mammal bones from these two settlements was 13,358 (including complete or partial skeletons).

Before examining the fractures from Tiel-Passewaaij, a short summary of the different types of fracture, the process of healing, the complications that may arise during healing, and the problems encountered when studying animal bone fractures are discussed.

**Figure 6.1**: map of The Netherlands indicating the location of Tiel-Passewaaij.

## Types of fractures

A fracture can be defined as "an incomplete or complete break in the continuity of a bone" (Lovell 1997, 141) and is usually the result of abnormal stress applied to a bone (Ortner 2003, 120). We can distinguish three basic kinds of fracture:

1. Acute fractures – resulting from either direct or indirect trauma;
2. Stress or fatigue fractures – caused by repetitive stress. The fracture line is usually perpendicular to the longitudinal axis of the bone and may resemble a transverse fracture. Stress fractures are usually not displaced and are often not visible on x-rays prior to callus formation (Lovell 1997, 144). Stress fractures often heal very well, and can be difficult to detect in archaeological samples (Ortner 2003, 125);
3. Pathological fractures – these occur when the bone structure has been affected by a local pathological process. Because the structure of the bone has become weaker, the bone is no longer able to withstand relatively normal biomechanical stress (Ortner 2003, 125). Underlying pathological conditions can be congenital, metabolic or infectious. Neoplasms can also weaken bone (Ortner 2003, 125). Osteoporosis commonly results in pathological fractures.

In acute fractures, a further distinction can be made between fractures caused by direct or indirect trauma. In direct trauma, there are three different fracture types. In a transverse fracture, the line of the break is perpendicular to the longitudinal axis of the bone and is caused by a relatively small force directed at a small area (Lovell 1997, 141). Penetrating fractures result from a large force delivered to a small area (Lovell 1997, 141) and are caused by sharp objects. In archaeological cases, it is very difficult to identify the object that caused the wound, unless the point has broken off and remained in the bone, although the absence of healing can show that the wound was severe enough to result in death. Finally, crush fractures occur in cancellous bone and result from direct force, which causes the bone to collapse (Lovell 1997, 142). Crush fractures can be caused by blunt trauma and are often found on the skull.

When a fracture occurs away from the location where the force was applied it is called indirect trauma (Lovell 1997, 142). Types of fractures resulting from indirect trauma are oblique, spiral, greenstick, impacted and avulsion fractures. In an oblique fracture, the fracture line angles across the longitudinal axis (Lovell 1997, 142). This type of fracture is caused by a combined angulated/rotated force. In a spiral fracture, the fracture line spirals around the longitudinal axis. This type of fracture is caused by rotational and downward loading stress on the longitudinal axis. In well-healed fractures, it is difficult to see the difference between oblique and spiral fractures (Lovell 1997, 142-3). Greenstick fractures are incomplete fractures and are common in non-adult bones, where the bone bends rather than breaks. Usually it is the convex side of the bone that breaks (Lovell 1997, 143). In an impacted fracture, the bone ends are forced into each other. In humans, this is often found in the proximal humerus as a result from a fall onto an outstretched hand (Lovell 1997, 143). An avulsion fracture occurs when a ligament or tendon is pulled away from its attachment to the bone and tears off a fragment of bone. Finally, a comminuted fracture occurs where the bone breaks into more than two fragments and this can be caused by either direct or indirect trauma (Lovell 1997, 143).

Apart from the types of fractures described above, we must also distinguish between open and closed fractures. A fracture where soft tissue and skin has also been injured, and where the fracture is exposed, is called an open or compound fracture. Open fractures are susceptible to infection.

## Healing

The healing of fractures generally follows a predictable sequence of events. First, a haematoma is formed at the site of the fracture. The haematoma is formed by blood flowing from vessels that have been torn by the fracture. The ends of the fractured bone die because of a lack of blood supply (Lovell 1997, 145, table 3). In the next stage, the haematoma is organised into a fibrous mass, uniting the fractured bone after about three weeks (Aufderheide and Rodriguez-Martin 1998, 21). In the third stage, primary bony callus forms within the fibrous mass (Ortner 2003, 126). The primary callus subsequently remodels into secondary callus and the woven bone transforms into lamellar bone. This provides a stronger union between the ends of the bone (Ortner 2003, 127). The final stage of healing is the remodelling of the bone to its original form (Lovell 1997, 145, table 3). The callus is reduced to the minimal amount that is necessary for biomechanical strength (Ortner 2003, 127-8).

Healing time is variable, but will be faster in cancellous bone than in cortical bone, and occurs twice as fast in children than in adults (Ortner 2003, 126, 128). Factors influencing the speed of healing are age, blood supply, fracture type, and skeletal element. Spiral and oblique fractures heal faster than transverse fractures (Lovell 1997, 145). Good health and nutritional state are extremely important if the healing is to be successful. Immobilisation of the injured bone aids healing, while mobility stimulates fibrous callus formation, which takes longer to heal. Infection or any other pathological process will delay healing (Aufderheide and Rodriguez-Martin 1998, 21). At least two weeks of healing are needed before the callus can be recognised in dry bones (Aufderheide and Rodriguez-Martin 1998, 23).

## Complications

Complications of acute fractures depend on the location and severity. Many complications can occur, although only the more common ones will be outlined here. First, inadequate fusion of the fracture can occur. Non-union can be diagnosed by rounded ends of the fractured bone, and a sealed marrow cavity; radio-logically, the bone ends show sclerosis. Non-union can result from infection, poor blood supply, deficiencies of vitamins or calcium, or lack of immobilisation of the fractured bone. If the fractured ends of the bone continue to move against each other, a false joint or pseudo-arthrosis may form (Lovell 1997, 147). Malunion occurs when a fracture heals with deformity, such as shortening or angulation. Misalignment and shortening of bones can cause other problems, such as osteoarthritis. A second complication is infection, although this is more likely to occur in an open fracture. Post-traumatic infection can be localised to the site of the fracture, but it can also spread through the bloodstream to other parts of the body (Lovell 1997, 146). Infection can show up as periostitis or osteomyelitis. Nerve damage is a further complication that can occur and can result in muscle atrophy; if the nerve loss is permanent, the bones will show disuse atrophy. If nerve damage involves loss of sensation at the fracture site, the individual will be less likely to immobilise the bone, and this will delay or prevent healing (Lovell 1997, 146). A relatively common complication of fractures is osteoarthritis. Osteoarthritis can be caused by joint fractures, or by abnormal biomechanical stress placed on the limb following a fracture. For example, the un-injured limb can be susceptible to osteoarthritis when it has to bear most of the body weight during the period when the injured limb is not used (Lovell 1997, 147). To distinguish between traumatically induced osteoarthritis and age-related osteoarthritis, two criteria are used: the presence of a fracture, and the absence of bilateral symmetry in the nature and degree of osteoarthritis (Ortner 2003, 157).

## Problems in studying fractures

Palaeopathologists can encounter several problems when studying fractures in archaeological material. First, the prevalence of fractures based on archaeological samples will always underestimate the true prevalence in past populations. Fractures in immature specimens may heal and remodel so completely that a fracture will not be recognised (Ortner 2003, 136); thus, fractures in non-adults will be underrepresented. Stress fractures are very hard to detect when healed, so they will also be underrepresented in archaeological samples. A second problem is that it is very difficult, and sometimes impossible, to distinguish between accidental fractures and fractures resulting from violence (Ortner 2003, 136). In order to make this distinction, it is necessary to understand the nature of the force causing the fracture. It is also difficult to distinguish between fractures that occurred just before death, and post-mortem fractures. Finally, in animal bone assemblages, we are usually dealing with fragments of individual bones and not complete skeletons. Consequently, complications of fractures may not be discernible since they may occur in other bones beside the fractured specimen. It also makes it difficult to identify animal abuse, since we cannot easily assess the distribution of fractures across individual skeletons (a key indicator of animal abuse; Teegen 2005), or the exact numbers of animals abused.

## Possibilities of studying fractures

Despite these problems, it is still worthwhile to study fractures. The prevalence and location of fractures in the skeleton are culturally influenced (Ortner 2003, 136), both in humans and in other animals. Domestic mammals have different functions in different societies, and some of these functions may predispose animals to fractures. Moreover, husbandry techniques can play an important role in fracture prevalence. Baker and Brothwell (1980, 91) offer the following research questions, specifically for palaeopathology in animals:

- is there a change in the frequency of fractures between time periods?
- is there a difference in the distribution of fractures for different species?

Additional questions that the analysis of fractures can also be used to explore include:

- what can fracture frequencies tell us about human-animal interaction, husbandry methods and animal function?
- does evidence exist for fracture treatment and what does this tell us about the value attributed to animals?

It is clear that in order to answer these questions more published data is required; the evidence presented below serves to provide such a contribution.

## Fractures from Tiel-Passewaaij

For this paper, 19 fractured mammal bones were studied. Two additional fractures were identified from the site, a horse (*Equus caballus* L., 1758) rib from the cremation cemetery in Tiel-Passewaaij and a chicken (*Gallus gallus* L., 1758) tibiotarsus; however, these were not considered for this study. The rib fracture was discarded because the animal bones from the cemetery are not comparable with the animal bones from the two settlements since many of the former have been cremated. The tibiotarsus was omitted because this paper focuses on mammals.

Tab. 6.1 provides a summary of the fracture evidence from Tiel-Passewaaij. All of the fractures were found in domestic mammals, they were all acute – no stress or pathological fractures were recognised – and no evidence for therapeutic intervention was apparent. The majority of fractures were found in dogs, and the most frequently fractured bones were ribs. Most rib fractures heal without any complications, because the bones are held rigidly in place. However, if rib fractures are caused by a massive injury, fragments can be displaced and penetrate the pleura, lungs or heart (Aufderheide and Rodríguez-Martín 1998, 25). Prior to the advent of modern veterinary care, this would usually have resulted in the death of the animal. However, the likelihood of distinguishing such unhealed ante-mortem fractures is small. Not all rib fractures will be discussed below; however, all non-rib fractures are described and illustrated.

| Location | Cattle | Horse | Pig | Dog | Total |
|---|---|---|---|---|---|
| Mandible | 1 | - | - | 1 | 2 |
| Ribs | 2 | 1 | 1 | 3 | 7 |
| Vertebrae | - | - | - | 1 | 1 |
| Humerus | - | - | - | 1 | 1 |
| Radius | - | - | - | 2 | 2 |
| Ulna | - | - | - | 1 | 1 |
| Tibia | - | - | - | 1 | 1 |
| Fibula | - | - | - | 1 | 1 |
| Metapodials | - | 1 | 1 | 1 | 3 |
| Total | 3 | 2 | 2 | 12 | 19 |

Table 6.1: location of fractures in the animal bones from Tiel-Passewaaij.

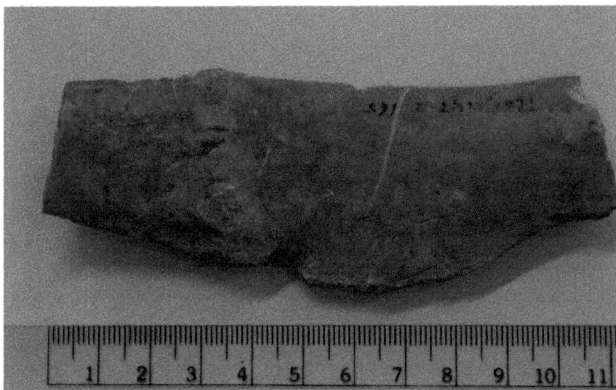

Figure 6.4: rib fracture in pig; pleural view (TLP OTW 36-295/24).

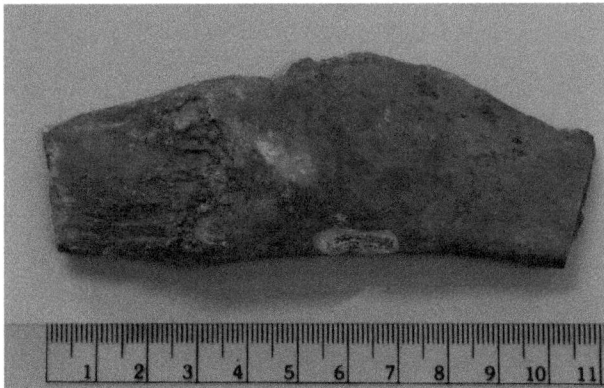

Figure 6.2: rib fracture in cattle; pleural view (TL 147-165).

Figure 6.5: rib fracture in pig (TLP OTW 36-295/24).

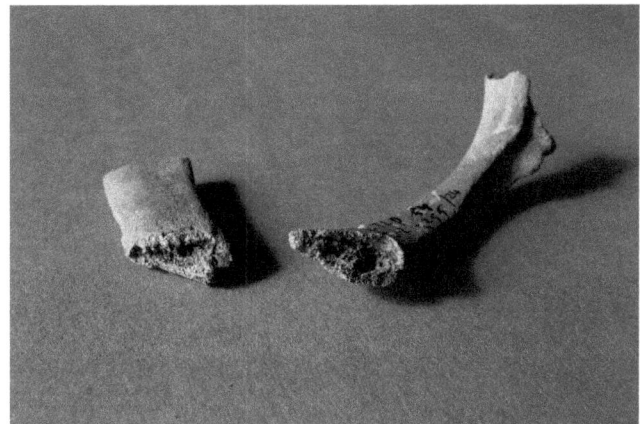

Figure 6.3: rib fracture in cattle; external view (TL 147-165).

Figs. 6.2 and 6.3 illustrate the united fracture of a cattle (*Bos taurus* L., 1758) rib. The callus in this specimen is irregular and the surface is rough on the external side but smoother on the pleural side. The fracture line is clearly visible, both on the bone and on an x-ray (as a radiolucent line).

The fracture of a pig (*Sus scrofa* L., 1758) rib illustrated in Figs. 6.4 and 6.5 has occurred not long before death. The fracture is located in the proximal third of the bone and has not united. There is some callus formation that is rough and porous.

A lumbar vertebra of a dog (*Canis familiaris* L., 1758) exhibited a fracture of the spinous process (Fig. 6.6), which healed with thick callus formation and had subsequently remodelled.

The right cattle mandible depicted in Figs. 6.7 and 6.8 reveals a fracture of the vertical ramus. Callus has formed and the two ends of the bone are in the process of fusing. The mandible has broken post-mortem at the fracture site. The callus is thick and part of the surface is porous, suggesting the fracture was relatively recent. The fact that the second molar has not erupted indicates the animal was not very old.

The left dog mandible in Fig. 6.9 has a fracture across the horizontal ramus, just in front of the first molar. The fracture has not yet fused and thus occurred shortly before death. Unfortunately, only the posterior end of the fracture site is present. There is some callus formation, which is porous and not remodelled. The mandible is part of an almost complete skeleton of an immature dog. The dog's age was estimated at 10-12 months according to Silver (1969, table A).

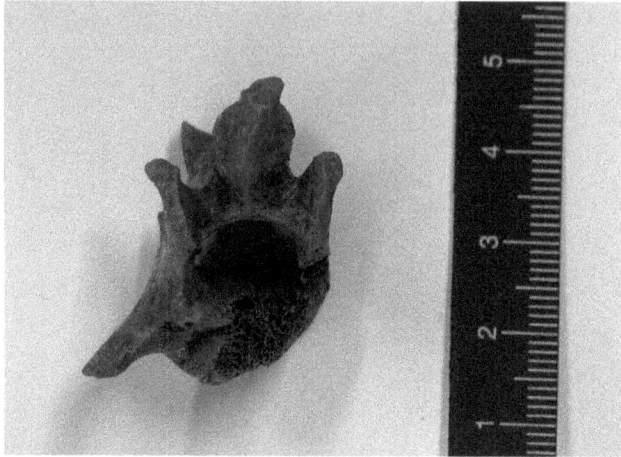

**Figure 6.6**: fracture in the spinous process of a dog lumbar vertebra; cranial view (TL 122-90).

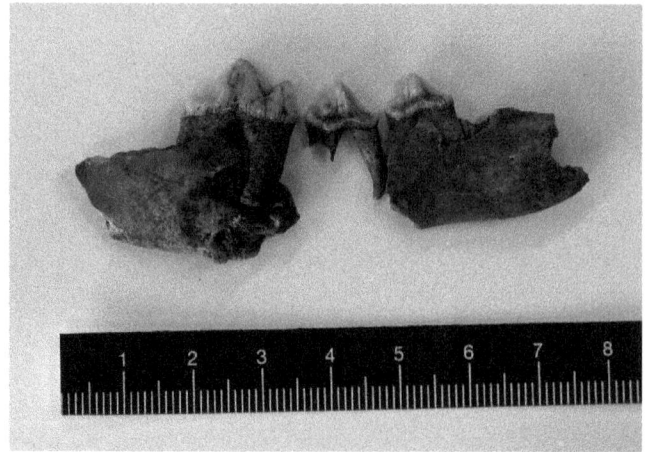

**Figure 6.9**: fracture in a dog mandible; medial view (TL 197-57).

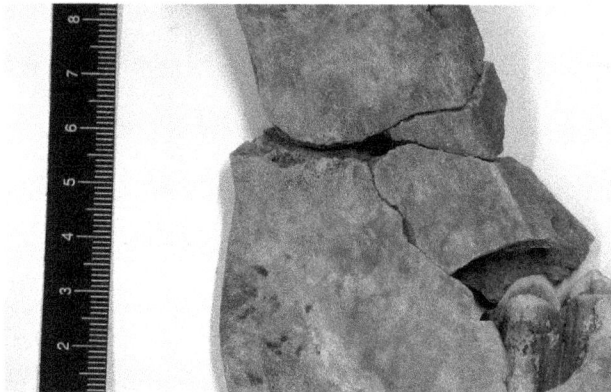

**Figure 6.7**: fracture in a cattle mandible; lateral view (TL 165-140).

Figs. 6.10-6.12 demonstrate a well-healed fracture of a right dog humerus. The fracture is located in the distal third of the diaphysis and appears to be oblique as evidenced in the x-ray as a radiolucent line. The two fractured ends of the bone have overlapped, leading to foreshortening; however, they have fused together and the callus has remodelled into normal bone, indicating that the dog survived long after it was injured. The greatest length of the humerus was measured, as well as that of the opposite, unaffected bone. The fractured humerus has a length of 151 mm, whereas the greatest length of the unaffected left humerus is 167 mm.

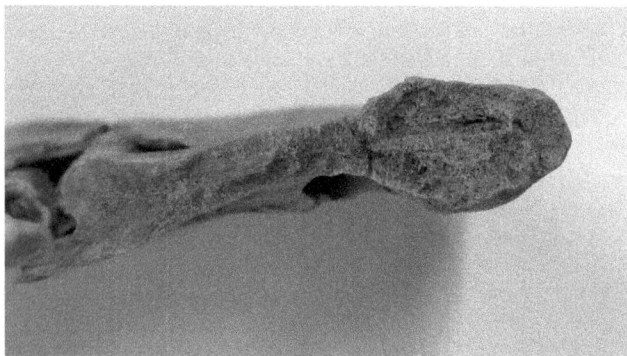

**Figure 6.8**: fracture in a cattle mandible (TL 165-140).

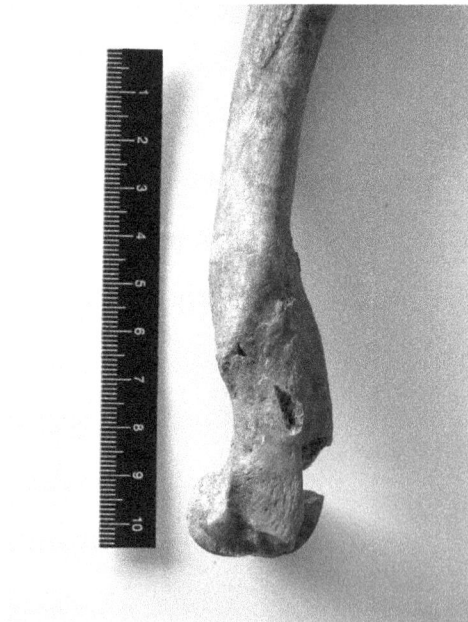

**Figure 6.10**: fracture in a dog humerus, medial view (TLP OTW 36-252).

44

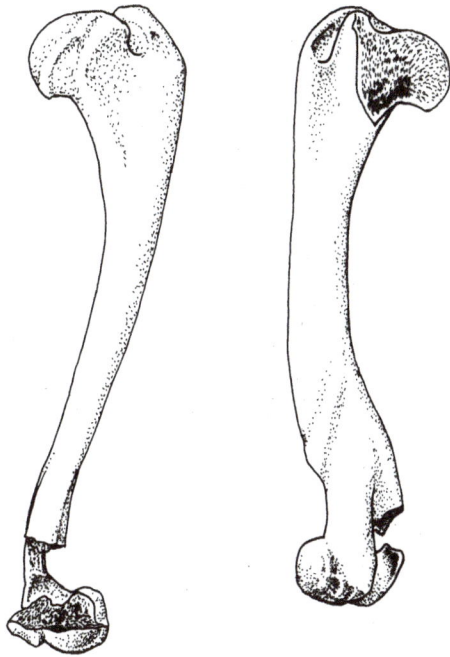

**Figure 6.11**: fractured dog humerus (right) united with shortening compared with the normal left humerus from the same individual (TLP OTW 36-252). Drawing by Mikko Kriek, ACVU-HBS.

A double fracture of a right dog radius and ulna was found in a complete dog skeleton (Figs. 6.13 and 6.14). The bone ends have fused with considerable distortion, uniting the radius and ulna together. The fracture line in the radius appears to be oblique in the x-ray. The two fragments are misaligned, both anterio-posteriorly and medio-laterally. The distal fragment is angulated to the lateral and posterior side. The medio-lateral view in the x-ray shows that the two ends do not touch at all. A large callus has formed around the bone ends, fusing the two parts. The greatest length of the radius could not be measured, because the distal epiphysis is missing. The anterio-posterior view of the ulna in the x-ray shows that the two fragments are closely aligned, but that the apposition is poor, although it is not possible to determine the type of fracture. The greatest length of the fractured ulna is 208 mm, while that of the unaffected left ulna is 215 mm. One of the left carpal bones (*os carpi radiale et intermedium*) exhibits eburnation, one of the signs of osteoarthritis. Considering the distortion of the radius and ulna, it is not surprising that the fracture had affected the joints below. This dog is a male adult individual with an average withers height of 60 cm (Harcourt 1974). Apart from the double fracture of the radius and ulna, the animal also suffered a rib fracture.

Fig. 6.15 illustrates a second radius fracture in a dog, occurring in the distal third of the bone. The fracture is well-healed: the callus has been remodelled and the marrow cavity had been restored. The bone is not complete, so no comment can be made regarding the extent of shortening or angulation.

A fracture of the right tibia and fibula occurred in the

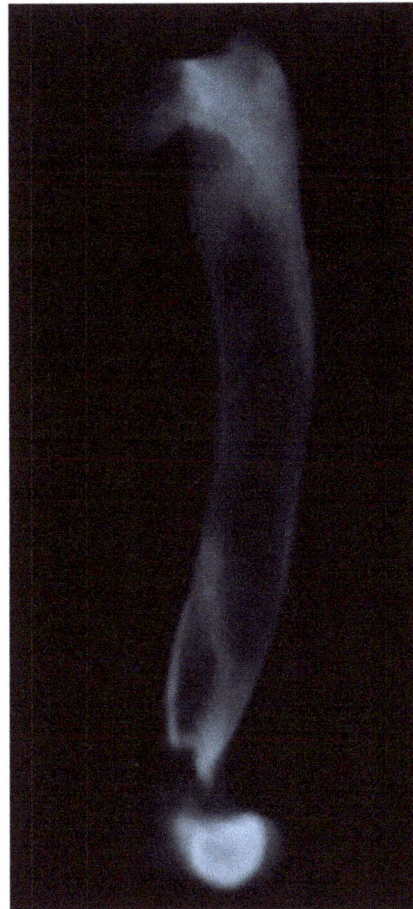

**Figure 6.12**: x-ray of a fractured dog humerus, lateral view (TLP OTW 36-252).

same dog that suffered the humerus fracture (Figs. 6.16 and 6.17). Because the bone was broken during excavation, and the other half was lost, the fracture type is not determinable. The fibula exhibits curvature, so it can be assumed that the realignment was not perfect and it is likely that the fractured tibia was shorter than the unaffected tibia. The tibia was fractured in the distal third of the diaphysis, and had fused, but with considerable distortion. A large callus has formed, and is well remodelled, although the marrow cavity had not reformed prior to the death of the animal. It is not possible to say whether the two fractures occurred at the same time. The ossification of a muscle attachment on the right femur of the same animal was also observed (Fig. 6.18). It is possible that this occurred because of the additional strain resulting from the animal holding its leg up during the healing process, or as a result of foreshortening.

While no fracture line was visible on x-rays of two right metatarsals (IV+V) of a dog, the presence of callus and the fusion of the two bones suggest the presence of an old fracture (Fig. 6.19). The bone has not completely remodelled and the surface is quite rough. It is possible that both bones were fractured simultaneously, around the mid-section of the diaphysis. The correct alignment can be explained by the presence of the other metatarsals, which would have acted as natural splints for the

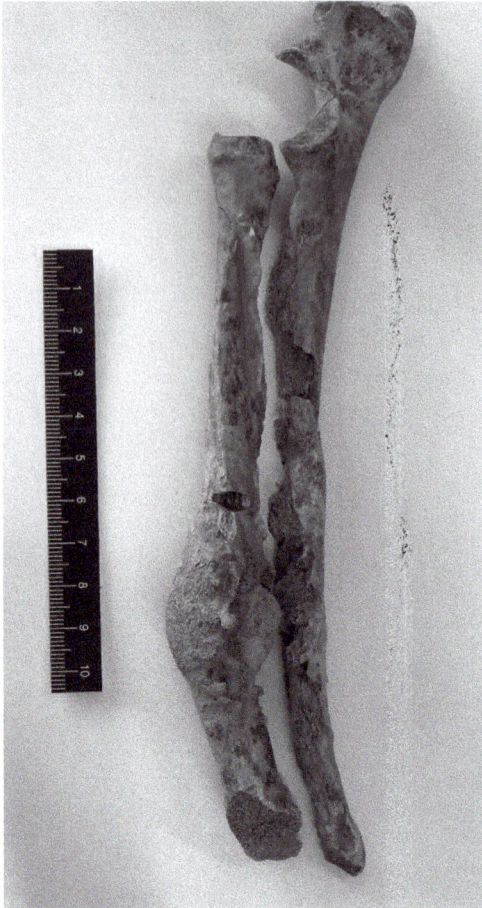

**Figure 6.13**: healed fracture in a dog radius and ulna; medial view (TL 165.150).

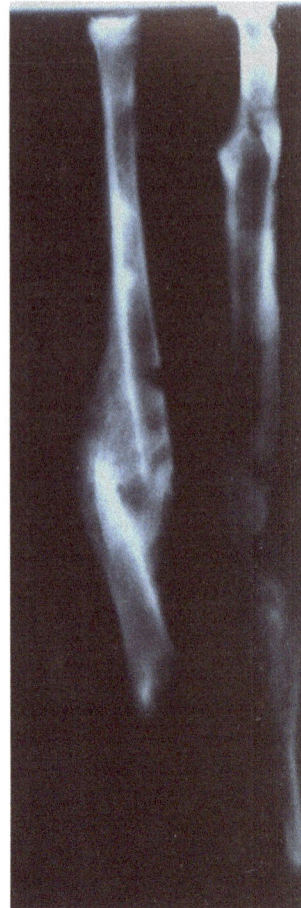

**Figure 6.14**: x-ray of a healed fracture in a dog radius and ulna; medial view (TL 165.150).

fractured bones. Well-healed metapodials without shortening or misalignment are commonly found in dogs (Udrescu and Van Neer 2005).

The right metacarpal of a horse is probably fractured (Fig. 6.20). An x-ray was taken, in which the fracture appears visible as a radiolucent line. The location of the fracture is close to the proximal end of the bone. The fracture has healed, fusing the second metacarpal to the third. In this case, the second metacarpal has acted as a natural splint (Udrescu and Van Neer 2005).

Finally, a right fifth metatarsus from a pig demonstrates a well-healed fracture (Fig. 6.21). A thick callus has formed around the fracture and has partly remodelled. The marrow cavity had not been restored before death.

## Discussion

If we compare the number of fractures with the total number of bone fragments, we can calculate prevalence for all domestic mammals, and for each separate species (Tab. 6.2).

The overall prevalence of fractures in the faunal assemblage from Tiel-Passewaaij is 0.14% for all domestic mammals. For dogs, the fracture prevalence is 0.93%, while for the other domestic species, prevalence

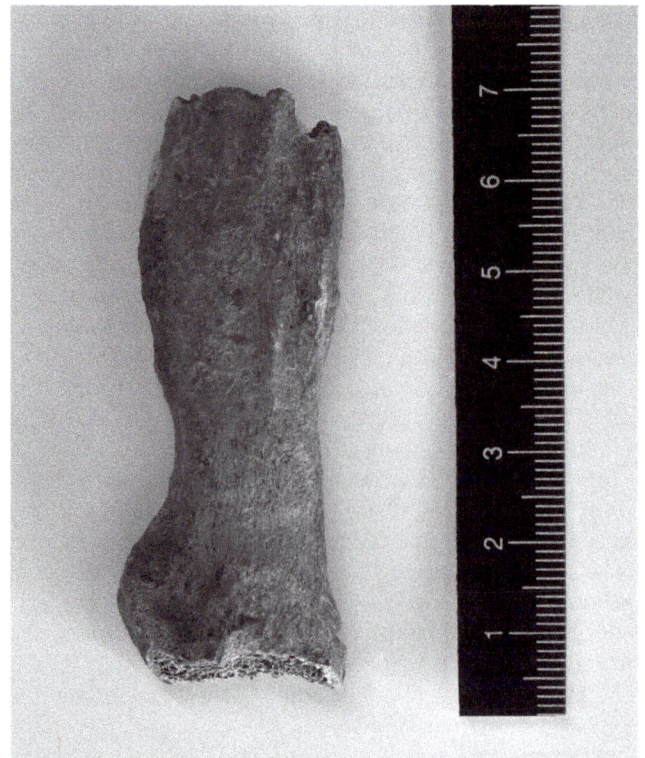

**Figure 6.15**: fracture in a dog radius; anterior view (TL 148-96).

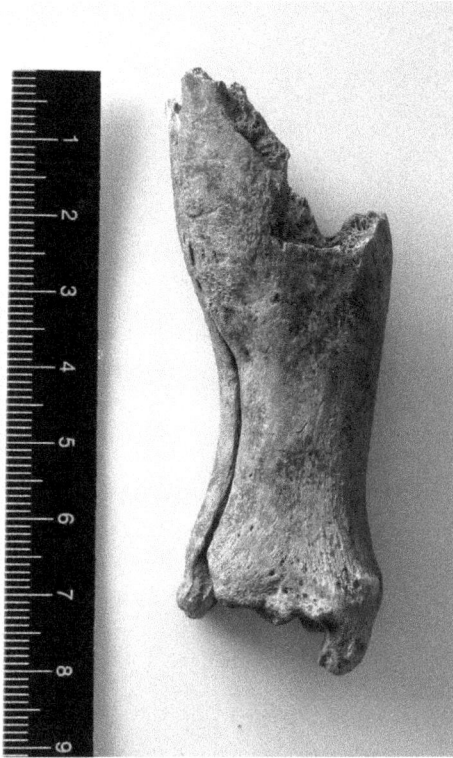

**Figure 6.16**: fracture in a dog tibia and fibula; anterior view (TLP OTW 36-252).

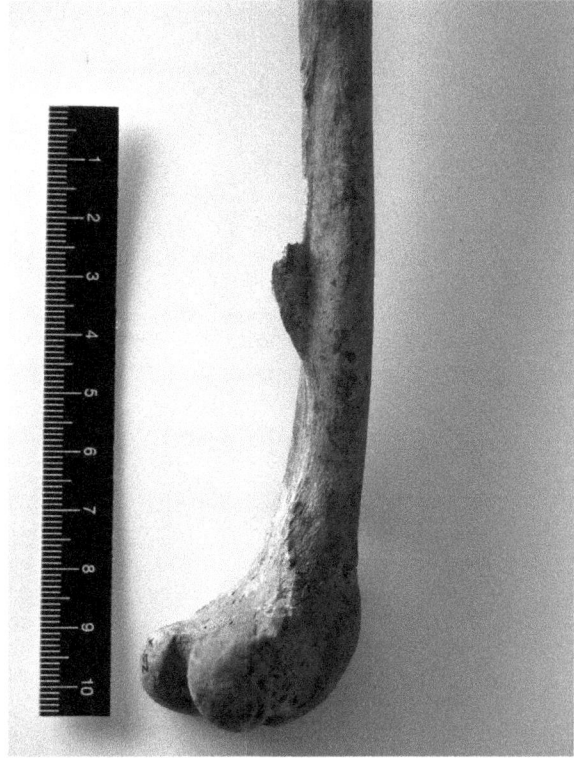

**Figure 6.18**: ossification in a dog femur; lateral view (TLP OTW 36-252).

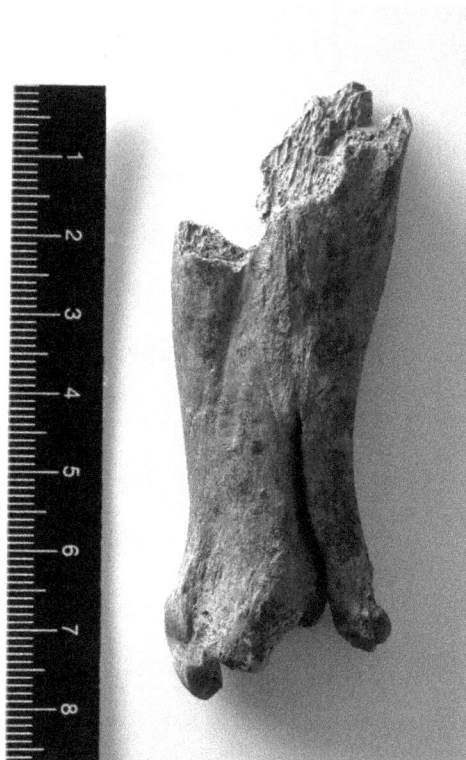

**Figure 6.17**: fracture in a dog tibia and fibula; posterior view (TLP OTW 36-252).

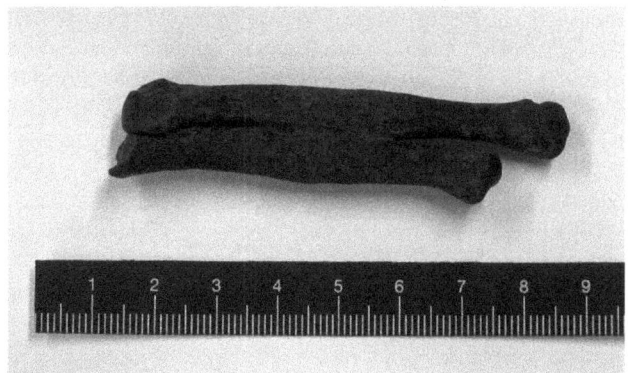

**Figure 6.19**: fractured dog metatarsals (TL 165-161).

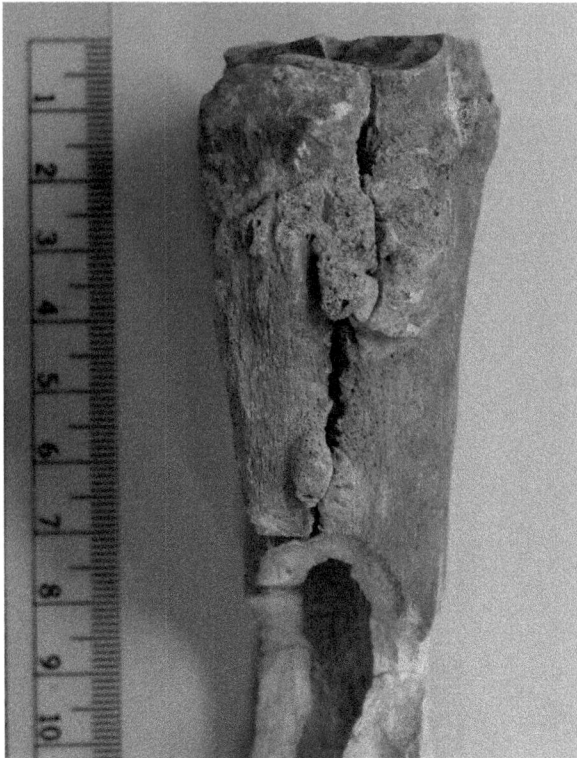

Figure 6.20: a probable fracture in a second horse metatarsal; lateral view (TL 179-171).

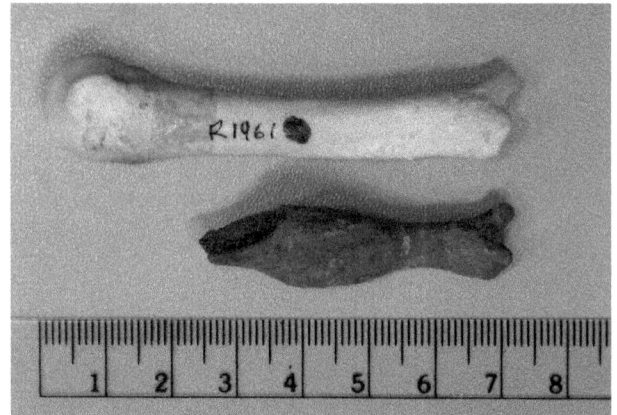

Figure 6.21: fracture in a pig fifth metatarsal, compared to a normal specimen (TL 170-92).

ranges from 0% for sheep (*Ovies aries* L., 1758) to 0.14% for pig. The prevalence for all domestic mammals together is a much higher number than the 'normal' frequency of 0.04% noted by Baker and Brothwell (1980, 91). The latter figure was obtained by combining results from their own survey with data published by Siegel (1976).

Why then is the fracture prevalence for Tiel-Passewaaij higher than that mentioned by Baker and Brothwell? The material included in the latter sample was from different periods; one possible conclusion therefore, is that fractures were more common in the Roman period.

Clearly, however, more systematic publication of fracture prevalence for sites from different periods is needed to answer this question (a point also emphasised by Thomas and Mainland 2005). What is a more interesting question at this moment is why the fracture prevalence is so much higher for dogs at Tiel-Passewaaij, compared to other domestic mammals.

One explanation is that fracture frequency is related to body size. Healed long bone fractures in cattle and horses are rare with only metapodial fracture known from the literature (Udrescu and Van Neer 2005). Fractures of the other long bones heal poorly and animals suffering long bone fractures in antiquity may have been slaughtered (as they often are today), in which case we would not find any evidence for their occurrence. Body size may explain the low fracture prevalence for horse and cattle, but if this was an important factor in fracture prevalence, we would expect to find similar frequencies in dogs, pigs and sheep. However, we have already seen that fracture prevalence is much higher in dogs than in other medium-sized mammals.

Dogs may be more susceptible to fractures than other mammals, because they live in closer proximity to humans. If this is indeed the case then we must consider maltreatment as a cause for the fractures. Teegen (2005) has discussed this issue, through the consideration of rib and vertebral fractures from medieval cities in Northern Germany. The dog bones in this sample showed a high frequency of rib and vertebral fractures.

| Species | Fractures | Total NISP | % fractures |
|---|---|---|---|
| Cattle | 3 | 5760 | 0.05 |
| Horse | 2 | 1950 | 0.1 |
| Pig | 2 | 1444 | 0.14 |
| Sheep | 0 | 2927 | 0 |
| Dog | 12 | 1277 | 0.93 |
| **Total** | **19** | **13,358** | **0.14** |

Table 6.2: prevalence of fractured bones by species at Tiel-Passewaaij.

One of the diagnostic traits for deducting abuse is the occurrence of multiple fractures in different stages of healing and Teegen found evidence for this in some partial skeletons. Other explanations for the fractures are mentioned: kicks from large animals, bite wounds from other dogs, and pathological fractures. However, Teegen (2005) believes that the presence of both rib and vertebral fractures in different stages of healing in individual animals is more suggestive of abuse by humans. Teegen and Wussow (2000) also discovered rib and vertebral fractures in nineteenth and early twentieth-century pigs and sheep, suggestive of kicking and beating by humans, and this interpretation is supported by documentary evidence for the use of pitchforks in handling these animals.

At Tiel-Passewaaij, both rib fractures and one fracture of the spinous process are present, but not in significant numbers. Multiple fractures in individuals are present: one dog suffered fractures of the humerus and tibia/ fibula, but because both fractures are well-healed, it is not possible to determine whether the injuries occurred simultaneously.

Another dog suffered a double fracture of the radius/ulna and a fractured rib, although again it is not possible to determine whether these fractures were the result of one or two traumatic events. In another dog, two ribs were fractured, probably at the same time.

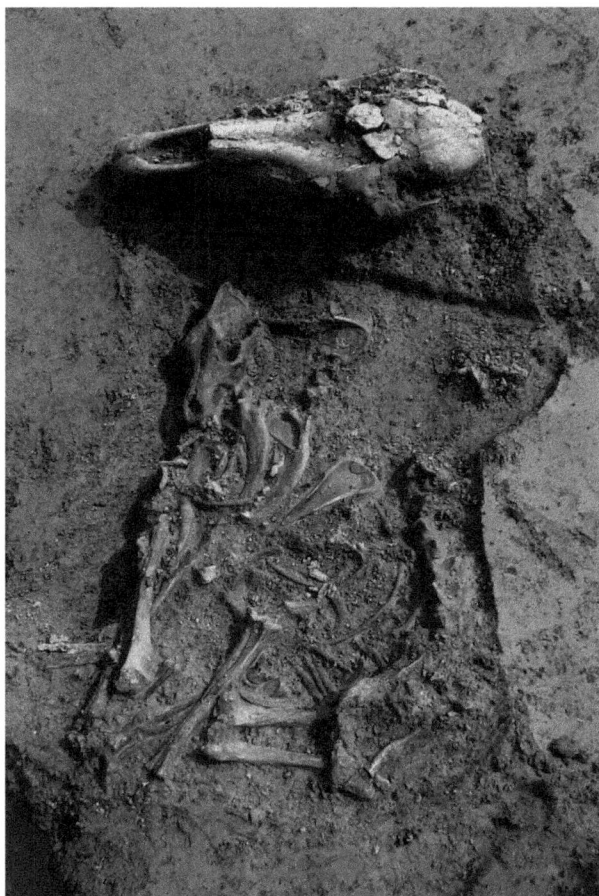

**Figure 6.22**: burial of a dog skeleton and a horse skull at Tiel-Passewaaij.

If we accept the hypothesis that fractures in dogs are the result of maltreatment by humans, we must consider why dogs were maltreated, and other domestic mammals were not. The spatial proximity of dogs to humans may provide one explanation. In this case, dogs were not intentionally treated worse than other animals; rather, the numbers of interactions (and opportunities for abuse) was greater. Another explanation is that dogs were not seen as valuable animals, and that they did not play an important role in society. Although we cannot discard this hypothesis, the careful burial of some dogs seems to contradict the idea of dogs being regarded in purely functional terms. The dog that suffered both a radius/ulna fracture and a rib fracture, for example, was carefully buried with a horse skull, in a ditch that surrounded a Late Roman part of the settlement (Fig. 6.22). Another dog was buried with the partial skeleton of a red deer, two dogs were buried on top of large pottery sherds (Fig. 6.23), and two more were buried in ditches surrounding houses. All these animals were buried with care and in a very deliberate manner. The location of the burials and the associated finds are not random. Dogs feature

prominently in the 'special animal deposits' that have been identified at Tiel-Passewaaij. Clearly, the roles dogs fulfilled were not just functional but also symbolic. However, as Thomas (2005) emphasises, the way in which animals are buried only tells us about their treatment in death, and not in life. Dogs may have been useful animals in life, and used in a symbolic manner after death, but this may not have saved them from maltreatment.

Another possibility is that the fractures in dogs are related to injuries sustained during work. Both hunting and herding can be dangerous activities, involving large, potentially aggressive animals. The association of a dog skeleton and a partial red deer skeleton is clearly suggestive of the use of dogs for hunting; however, wild mammals are rare at Tiel-Passewaaij, so hunting was probably not a very common activity. Therefore, it is unlikely that hunting injuries would be an important cause of fractures in dogs. Livestock, on the other hand, played a vital role in the rural economy and animals were probably traded with the Roman army or the city. Dogs could have been valuable companions, helping transport cattle to markets, for instance, or rounding up animals from the fields. This would predispose them to kicks from cattle and horses, which could easily result in fractures. Even the small dog with the broken mandible could have been a herding dog: nowadays, several breeds of small dogs still exist that were originally developed for herding cattle, such as the Welsh corgi, the Lancashire Heeler and the Swedish Vallhund.

**Figure 6.23**: burial of a dog on top of large pottery sherds at Tiel-Passewaaij.

A final possibility we must consider is intra-species aggression. Although fractures can, in theory, be a result of aggression between dogs, this does not seem to be plausible explanation in this case. Bite wounds in dogs are often directed at the head and shoulder region, rather than the extremities (Baranyiová *et al.* 2003, 58-59). Further-more, none of the fractures seem to be caused by puncture wounds.

## Conclusions

Currently, it appears that the high fracture prevalence for dogs in Roman Tiel-Passewaaij is either the result of maltreatment by humans or kicks from large animals. More research into the location of fractures in known herding dogs as well as known abused dogs could perhaps help identify the cause of the dog fractures in Tiel-Passewaaij (Thomas and Mainland 2005) and could tell us more about the function of dogs and human attitudes to them.

In addition to research into modern fracture rates, more work also needs to be done using archaeological material. Patterns of fractures need to be studied, and for this large samples of animal bones are needed. Fractures should be recorded and published systematically, preferably with illustrations. It is important not just to describe fractures in reports on animal bones, but also to publish prevalence rates of fractures for different species and different time periods. Only by gathering a large amount of data, will we be able to make any more meaningful conclusions in the future.

## Acknowledgements

I would like to thank the following people and organisations for their support: NWO (Netherlands Organisation for Scientific Research) who financed the Ph.D. research into the animal bones from Tiel-Passewaaij; Prof. Dr G.J.R. Maat of Barge's Anthropologica (Leiden University Medical Centre) for some helpful advice, and for introducing me to Prof. Watt; Prof. Dr I. Watt of Leiden University Medical Centre, who kindly arranged for x-rays to be taken of some of the fractured bones from Tiel-Passewaaij; and finally, an anonymous referee for their helpful comments on this text.

## Bibliography

Aufderheide, A. C. and Rodríguez-Martín, C. 1998. *The Cambridge Encyclopedia of Human Paleopathology*. Cambridge: Cambridge University Press.

Baker, J. and Brothwell, D. 1980. *Animal Diseases in Archaeology*. London: Academic Press.

Baranyiová, E., Holub, A., Martiníkova, M., Nečas, A. and Zatloukal, J. 2003. Epidemiology of intra-species bite wounds in dogs in the Czech Republic. *Acta Veterinaria Brno* 72, 55-62.

Harcourt, R. A. 1974. The dog in prehistoric and early historic Britain. *Journal of Archaeological Science* 1, 151-75.

Lovell, N. 1997. Trauma analysis in paleopathology. *Yearbook of Physical Anthropology* 40, 139-170.

Ortner, D. J. 2003. *Identification of Pathological Conditions in Human Skeletal Remains* (second edition). San Diego, USA: Academic Press.

Siegel, J. 1976. Animal palaeopathology: possibilities and problems. *Journal of Archaeological Science* 3, 349-384.

Silver, I. A. 1969. The ageing of domestic animals, pp. 283-302, in Brothwell, D. and Higgs, E. S. (eds), *Science in Archaeology* (second edition). London: Thames and Hudson.

Teegen, W. R. 2005. Rib and vertebral fractures in medieval dogs from Haithabu, Starigard and Schleswig, pp. 34-38, in Davies, J., Fabiš, M., Mainland, I., Richards, M. and Thomas, R. (eds), *Diet and Health in Past Animal Populations: Current Research and Future Directions*. Oxford: Oxbow.

Teegen, W. R. and Wussow, J. 2000. Maltreatment of animals in the late 19th and early 20th century AD? Evidence from the Julius-Kühn Collection, University of Halle-Wittenberg (Germany). Poster presented at the European Meeting of the Paleopathology Association at Chieti, September 2000.

Thomas, R. 2005. Perceptions versus reality: changing attitudes towards pets in medieval and post-medieval England, pp. 95-104, in Pluskowski, A. (ed.), Just Skin and Bones? *New Perspectives on Human-Animal Relations in the Historic Past* Oxford: British Archaeological Reports International Series 1410.

Thomas, R. and Mainland, I. 2005. Introduction: animal diet and health – current perspectives and future directions, pp. 1-7 in, Davies, J., Fabiš, M., Mainland, I., Richards, M. and Thomas, R. (eds), *Diet and Health in Past Animal Populations: Current Research and Future Directions*. Oxford: Oxbow.

Udrescu, M. and Van Neer, W. 2005. Looking for human therapeutic intervention in the healing of fractures of domestic animals, pp. 24-33, in Davies, J., Fabiš, M., Mainland, I., Richards, M. and Thomas, R. (eds), *Diet and Health in Past Animal Populations: Current Research and Future Directions*. Oxford: Oxbow.

## Author's affiliation

Archeological Centre
VU University Amsterdam
Faculty of Arts
De Boelelaan 1105
1081HV Amsterdam
The Netherlands

# 7. Pathology in horses from a Roman cemetery in Budapest, Hungary

Kyra Lyublyanovics

## Abstract

*Skeletons of two horses (*Equus caballus *L., 1758) and parts of a wagon came to light during excavations of a second century AD cemetery on Bécsi Road, Budapest, Hungary. The wagon was burned and the two mares were probably sacrificed. There a few other examples of this Iron Age custom throughout the former Roman province of Pannonia; such graves have been found across the whole territory of the tribe of the Eravisci. However, these individuals displayed pathological changes that might possibly relate to the way in which they were used and kept and shed further light upon the selection policy of animals for sacrifice. Until now, pathological individuals have only been reported in one other similar context, making these horses an almost unique find.*

## Introduction

During excavation works at Bécsi Road, Budapest, Hungary, in 2003, the remains of a second century AD cemetery were found. Two complete horse (*Equus caballus* L., 1758) skeletons came to light during the excavations; the animals were buried together with the parts of a wagon, probably as a sacrifice. The wagon was burned before it was buried. Originally both skeletons were complete; however, a metatarsal and two phalanges were lost during recovery.

These horses offered a great opportunity to examine pathological changes caused by their upkeep and use. This material is also interesting because the pathological alterations observed in the two skeletons are similar but must be considered differently. Furthermore, these horses show how important it is to consider the muscular system, not just the bones, in any interpretation of pathology.

## Materials and methods

The remains belong to the collection of the Aquincum Museum, Budapest. The terminology used in this paper and the methodology of the measurements follow von den Driesch (1976). The estimation of the withers height is based on the methods of Kiesewalter (1888).

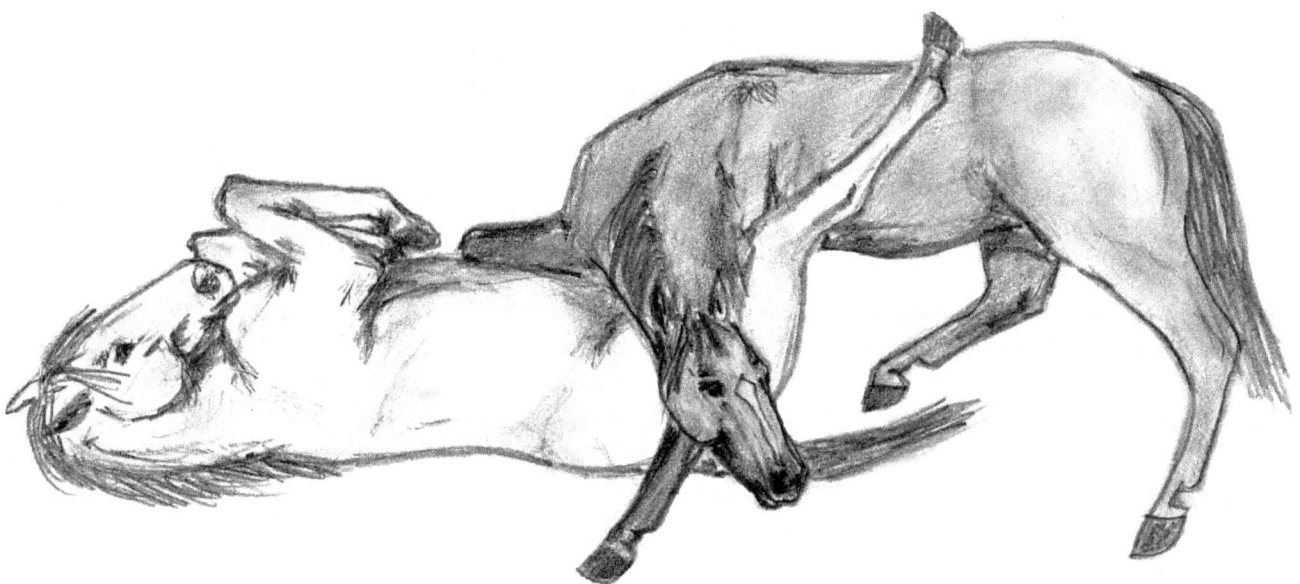

**Figure 7.1**: reconstructed position of the horses (drawn by the author).

51

## The remains

The horses were found in anatomical position, lying on each other *in situ*. The first individual lay on its side, the other on its belly, with forelegs bent and back legs stretched. One back leg of the horse that lay underneath was placed on the body of the other individual. The position shows that the animals were struck down beside the pit and they were moving after they fell into it (their reconstructed position is depicted in Fig. 7.1).

The method of killing is not clear; there was no evidence for this on the bones. The cervical vertebrae were largely intact, although the atlas was fragmented. It was impossible to distinguish peri-mortem fractures from post-mortem fractures; consequently, it was not possible to determine whether any injuries were sustained on the vertebrae.

It is possible that the animals were led to the pit on a cord or on a halter made of material that decomposes easily. Only one iron ring that was likely to have derived from a harness was found, located on the ribs of the underlying horse. The wagon lay close to the animals and its iron parts caused green and brown stains on the bones. The foreleg bones of the underlying individual were burned *in situ*.

Based on the shape of the pelvis and the lack of canines, both animals were probably mares. We can determine the age on the basis of the incisors and the ossification of the epiphyses. The older individual was 18 years old, while the other was at least 42 months old.

## The younger individual

The younger horse was discovered in poor condition; the bones were fragmented, although the skeleton was complete. Only some fragments of the frontal bone, the zygomaticus, the occipital condyle, the maxilla and the petrosal bone could be determined from the skull. There was a green stain on the right tibia caused by the iron parts of the wagon.

The skeleton was very robust. The long bones were wide and well grown; the metatarsal bones were 15% longer than the metacarpals. The bones were in the last growing phase when the animal died, and the line between the diaphysis and the epiphysis is discernible. Strong tuberosities were characteristic of all the long bones. The phalanges show a strong pastern. The withers height, estimated from the greatest lengths of the long bones (Tab. 7.1), was 149 cm, suggesting that the animal was of Roman, rather than Iron Age, origin: indeed, it would have been a respectable size for Roman horses of 'military type' (Bökönyi 1974, 263-267; 1989, 53; Peters 1998, 152-153). During the Imperial Period, animals of Greek, Persian and Thracian origin were used by the Roman army. Examples of these large individuals have been found over the whole area of present-day Hungary, beside the small-sized horse population of the local inhabitants (Bökönyi 1974, 263).

A number of pathological lesions were observed on the bones. They do not seem to be very significant; however, their presence is curious because they are similar to pathologies more typical of very old animals, and they are located across the whole of the skeleton. Most of them are 1-5 mm long exostoses. On the long bones they comprise island-like protuberances, with clear contours. There are compact exostoses with uneven surface on the caudal sides of the radii (Fig. 7.2), and on four proximal phalanges and the distal phalanges of the forelimb; the back legs were normal. On the right mandible a callus-like thickening was present. The ramus of both mandibles is thickened and there are small pin-like exostoses on the medial side (Figs. 7.3 and 7.4).

**Figure 7.2**: compact exostosis on the caudal side of the right radius of the younger individual.

On the spine no pathological protrusions were evident; the thoracic and lumbar vertebrae were intact and the ventral crest was strong. There were some small exostoses on the sacrum which are not necessarily pathological. The epiphyses had become detached from some of the vertebrae. The scapulae and ribs were normal. On the ischial spine of the pelvis (on both sides) a pathological growth was observed, with uneven and sharp surfaces (Fig. 7.5). The skull was completely fragmented but no pathological phenomena were discovered on the fragments.

## The older individual

The older horse was discovered in better condition. This 18-year old mare was a little smaller (the withers height estimated from the long bones was 147 cm), with shorter and more robust legs (Tab. 7.2). Only the *incisivum* and some mandibular fragments were pre-

Figure 7.3: exostoses on the medial side of the left mandible of the younger individual.

Figure 7.4: thickening on the ramus of the right mandible of the younger individual; caudal view.

| | Left | | Right | |
|---|---|---|---|---|
| Bone | GL (mm) | WH (cm) | GL (mm) | WH (cm) |
| Humerus | 300 | 150.2 | 305 | 152.5 |
| Radius | 340 | 147.6 | 345 | 149.7 |
| Metacarpus | 235 | 147.4 | - | - |
| Femur | 400 | 140.4 | 400 | 140.4 |
| Tibia | 374 | 163.0 | 379 | 165.2 |
| Metatarsus | 275 | 143.9 | 275 | 143.9 |
| Calculated withers height based on the long bones: 149.4 cm | | | | |

Table 7.1: greatest length (GL) of the bones of the younger individual and the estimated withers height (WH).

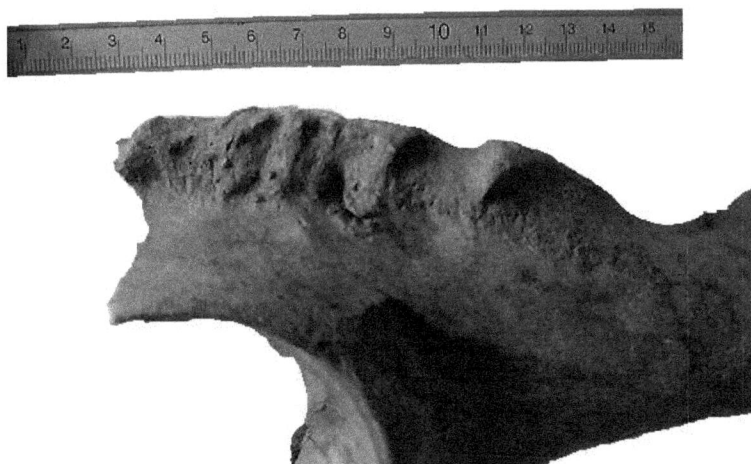

Figure 7.5: exostoses on the right pelvis of the younger individual.

| Bone | Left GL (mm) | Left WH (cm) | Right GL (mm) | Right WH (cm) |
|---|---|---|---|---|
| Humerus | 285 | 141.5 | 290 | 145.0 |
| Radius | 335 | 145.4 | 331 | 143.6 |
| Metacarpus | 230 | 144.2 | - | - |
| Femur | - | - | 380 | 133.4 |
| Tibia | 352 | 153.5 | 350 | 152.6 |
| Metatarsus | - | - | 273 | 142.8 |
| Calculated withers height based on the long bones: 147.4 cm | | | | |

Table 7.2: greatest length (GL) of the bones of the older individual and the estimated withers height (WH).

-served from the skull; the incisors were very worn. The bones of the legs were measurable, however, the bones of the forelegs were burned and the metacarpus and the phalanges of the right foreleg were fragmented.

Similar, but not necessarily identical, pathological phenomena were observed on the bones of this individual, but in this case we can consider them to be a consequence of age. On the distal condyle of the right humerus a number of small exostoses were evident (Fig. 7.6) and the muscle attachment surfaces of the radii, femora and tibiae were rugose. On the proximal phalanges changes similar to the younger individual were apparent: small protrusions on both the medial and lateral sides (Fig. 7.7).

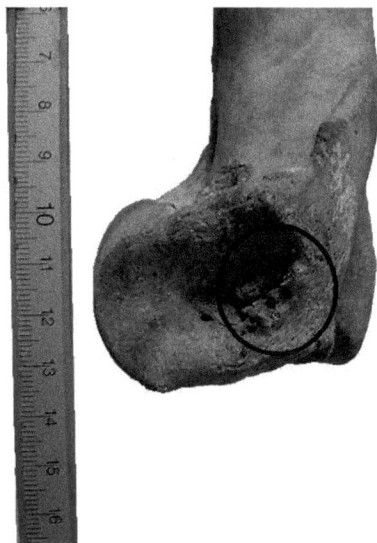

Figure 7.6: small exostoses on the distal condyle of the right humerus of the older individual.

In addition, the last two lumbar vertebrae were fused. The muscle attachment surfaces were pronounced on the alae of the sacrum. On the ischial spine of the pelvis an overgrowth was evident, but it was not as developed as in the other mare.

## Discussion

Graves with horses and wagons have been found on several sites in the area of present-day Hungary: Dunapentele, Enying, Etyek, Káloz, Környe, Nagylók, Nagytétény, Óbuda, Petrovina, Poljanec, Sárszentmiklós, Szomor, Vajta, Velence, Zomba, Zsámbék, Inota, Baláca and Kozármisleny (Lányi and Mócsy 1990, 245). These sites are located in the former habitation area of the Iron Age tribe of Eravisci, and are mostly dated to the second century AD. In this region, Roman period tombstones with depictions of wagons pulled by horses, asses or mules are also frequent (Sági 1945, 217; 1951, 74). The custom of burying wagon and horses seems to have been characteristic of the Romanised aristocracy of the Eravisci (Lányi and Mócsy 1990, 246). The idea of the dead travelling to the underworld on a wagon was commonly-held by the local Iron Age inhabitants of Pannonia and might have provided the basis for the Roman custom (Lányi and Mócsy 1990, 245). Until now, however, no such finds have been discovered in the area of Roman Aquincum (Facsády 2004, 27).

Figure 7.7: proximal phalanx of the left hind leg of the older individual.

Unfortunately, most of the sites detailed above were excavated in the first half of the 20th century and thus the animal bones were either not properly analysed or not even collected. At Inota, two stallions and two other individuals (probably also stallions) were recorded, while in Kozármisleny the remains of two mares were found (Bökönyi 1981, 47; 1989, 54). Graves with wagons are normally supposed to include three horses (one pack animal and two for pulling the wagon; Palágyi and Nagy 2000, 81); however, there is no clear evidence for this practice, and we have some graves with wagons but no horses (Ratimorská 1982, 255-256). The wagon was sometimes burned but the only other site at which the animals were burned was Baláca (Palágyi and Nagy 2000, 82). The horses and the parts of the wagon were usually buried in different pits (Palágyi and Nagy 2000, 149). Sometimes the spears with which the animals were killed were also discovered in the pit (Marosi 1935, 213; Palágyi and Nagy 2000, 82).

Horses with pathological phenomena have only come to light once from such a context. At Inota, skeletons of four adult individuals were found; two of which were identified as stallions. These individuals were of a similar size to the horses from Bécsi Road; their estimated withers height was between 141.4 and 147.6cm. The second, third, fourth and fifth lumbar vertebrae of one individual were fused, with gigantic exostoses on the third and fourth vertebrae. The zooarchaeologist Sándor Bökönyi identified this phenomenon as *arthritis chronica* (Bökönyi 1981, 47-48).

In order to interpret the pathological changes observed in the horses from Bécsi Road it is necessary to consider all possible diagnoses. However, it is difficult to identify a disease only on the basis of symptoms observed on the bones. Exostoses are mostly caused by inflammation or the abnormal statics of the skeleton (Sályi 1965, 466). In the case of the younger individual, the pathology of the radii and phalanges may have resulted from overloading of the legs or mechanical damage (Péter Sótonyi pers. comm.). If the *musculus extensor digitorum* had become overloaded, this would have resulted in the attachment surfaces of this muscle became more pronounced; it is perhaps for this reason that protrusions were observed on the radii. These pathologies can result from working as a pack animal or from constant concussion against a hard surface. The fact that only the forelimbs were affected can be explained by the fact that they carry 58.5% of the weight of the trunk (Fehér 2000, 239). Another possible explanation is that the horse was frequently tethered and the abnormal position of the forelegs caused pathological overgrowth, first on the proximal phalanges and then on the radii. However, we have no supporting clinical evidence to test this hypothesis – tethers are scarcely used nowadays, and thus the modern equine veterinary literature makes little mention of the skeletal consequences of this practice. The pathology of the pelvis can also relate to the overloading of the articulation. Before death, the motion was probably hard and painful for the animal.

The pathology of the mandible may have been caused by a hard blow. The bone split as a consequence, and even though it was not exposed to infection, it caused thickening (through callus formation) and exostoses formation. The exostoses are quite small but we can consider them as serious, because the little protrusions penetrated into the muscle (*musculus pterygoideus lateralis*), and probably caused pain and made the animal almost unable to chew. It is likely that in the last phase of the disease the horse was unsuited for work because of the painful motion and chewing.

In the case of the older horse, the observed pathological lesions were probably due to senescence. In heavy horses, such changes are not necessarily pathological and the over-growths observed are not so pronounced. Fusion of the vertebrae and *spondylitis deformans* is frequent in horses of advanced age (Karsai 1982, 644). The exostoses of the proximal phalanges may also be rooted in hard work in this case but this cannot be stated with any certainty.

## Conclusions

There are various explanations for the observed pathological phenomena; however, it is difficult to identify the exact cause of these changes. It seems logical that the animals were sacrificed because of their inability for work, which was a consequence of disease and age. In the case of the younger individual, it can be stated that the horse could not be used as a working animal anymore. It is possible, therefore, that only those animals that were not fit were considered for sacrifice because they would have represented a less substantial economic loss to the attendant community.

However, this custom was characteristic for the aristocracy of the Eravisci (Lányi and Mócsy 1990, 246), and limitation of economic loss may have been a lesser concern. Horses found in such contexts almost never display pathological changes (or at least, it is not mentioned in the old publications). The deceased person's affection towards these animals is also a possible explanation for the choice. The size and stature of the horses indicate their Roman origin, which means that they may have been valuable, imported animals in the eyes of the local inhabitants. Nevertheless, we need more, well-documented and analysed graves to support any theories about this custom. Reanalysis of curated skeletons excavated in the first half of the twentieth century may offer further valuable data.

## Acknowledgements

I would like to thank Annamária Facsády, the excavator of the cemetery, for allowing me to use the data, and Péter Sótonyi, professor in veterinary anatomy at the St. Stephan University in Budapest, for his comments on the pathology of the bones.

## Bibliography

Bökönyi, S. 1974. *History of Domestic Mammals in Central and Eastern Europe*. Budapest: Akademiai Kiado.

Bökönyi. S. 1981. Untersuchung der Tierknochenfunde des römerzeitlichen Gräberfeldes von Várpalota-Inota. *Alba Regia* 19, 46-52.

Bökönyi, S. 1989. Die Pferdeskelette des römischen Wagengrabes von Kozármisleny (Anhang) 51-62. In Kiss, A. (ed.), *Die römerzeitliche Wagengrab von Kozármisleny, Ungarn, Kom. Baranya.* Budapest: Magyar Nemzeti Múzeum.

Driesch, A. von den 1976. *Guide to the Measurement of Animal Bone from Archaeological Sites.* Harvard: Peabody Museum of Archaeology and Ethnology.

Facsády, R. A. 2004. Temetőfeltárás a Bécsi úton. *Aquincumi Füzetek* 10, 21-29.

Fehér, Gy. 2000. *A háziállatok funkcionális anatómiája.* Budapest: Mezőgazda.

Karsai, F. (ed.) 1982. *Állatorvo; si kórélettan.* Budapest: Mezőgazdasági Kiadó.

Kiesewalter, L. 1888. *Skelettmessungen am Pferde.* Leipzig: Inaugural Dissertation.

Lányi, V. and Mócsy, A. 1990. Temetkezés és halott-kultusz, pp. 243-253, in Mócsy. A. and Fitz. J. (eds), *Pannonia régészeti kézikönyve.* Budapest: Akadémiai Kiadó.

Marosi, A. 1935. A Székesfehérvári Múzeum római kocsilelete Kálozról. *Archaeologiai Értesítő* 48, 213-216.

Palágyi, S. K. and Nagy. L. 2000. *Római kori halomsírok a Dunántúlon.* Veszprém: Szerk.

Peters, J. 1998. *Römische Tierhaltung und Tierzucht.* Passauer Universitäts Schriften für Archäologie, Bd 5: Verlag Marie Leidorf.

Ratimorská, P. 1982. A környei 2. számú római kocsilelet. *Arcaheologiai Értesítő* 109, 255-275.

Sági, K. 1945. Kocsiábrázolások Pannonia császárkori szepulkrális vonatkozású kőemlékein. *Archaeologiai Értesítő* 5-6, 214-231.

Sági, K. 1951. Adatok a pannoniai császárkori kocsitemetkezések ethnicumának kérdéséhez. *Archaeologiai Értesítő* 78, 72-78.

Sályi, Gy. 1965. *Háziállatok részletes kórbonctana.* Budapest: Mezőgazdasági Kiadó.

## Author's affiliation

Institute of Archaeological Sciences
Loránd Eötvös University
Múzeum körút 4/B
1088 Budapest
Hungary

# 8. Animal diseases at a Celtic-Roman village in Hungary

Márta Daróczi-Szabó

## Abstract

*The large faunal assemblage from the predominantly Roman settlement (second century AD to the second half of the third century AD) of Balatonlelle-Kenderföld in western Hungary yielded a variety of animal bones with pathological lesions. Fifty-four of the over 9,000 identifiable specimens have been evaluated from a palaeopathological perspective. Morphological symptoms of trauma, inflammation, oral pathology and even tuberculosis could be identified on the bones of cattle (*Bos taurus *L., 1758), pig (*Sus scrofa *L., 1758) and dog (*Canis familiaris *L., 1758) respectively. In spite of their major contribution to this assemblage, the sample of small ruminants yielded no pathological bone specimens. The relationship between sampling and the manifestation of disease in archaeozoological assemblages is considered.*

## Introduction

At present, archaeological excavations precede the construction of new motorways in many areas of Hungary as a legal requirement. One of the sites thus recovered was Balatonlelle-Kenderföld on the shore of Lake Balaton in south-western Hungary (Fig. 8.1). A research area of approximately 39,000 square meters was uncovered during 2002 under the direction of Tibor Marton and Gábor Serlegi. The excavated material represents a broad time interval spanning the late Neolithic to the Roman period, which corresponds to the mid-third century AD in the province of Pannonia. Some 14,000 animal remains were recovered from this site, two thirds of which were identifiable to species (NISP=9291). Given the relative rarity of pathological specimens in archaeozoological assemblages, this large sample offered a good opportunity to review several cases and attempt their diagnosis.

suggest that the village was inhabited by Romanised Celts even during the Roman period, up to the middle of the third century AD.

**Figure 8.1**: location map.

## Chronological context

The chronological distribution of animal bones is shown in Fig. 8.2. The majority of prehistoric bones (late Neolithic and Bronze Age), which were only found in small numbers, could not be precisely dated and thus have limited interpretative potential. No pathological phenomena were observed in this, the smallest sample of the faunal assemblage. The contribution of bones from the Celtic period was 7%, almost negligible compared to the 87% of the material originating from the Roman period. Pork was a characteristic feature of meat consumption during the Celtic period. This is clearly shown even in this relatively small assemblage, in which pig remains comprise 35% of the total NISP. Since most pigs tend to be slaughtered young, osteopathies seldom develop in these animals. Nevertheless, even the small Celtic component is important since the archaeologists presume that this rural settlement was inhabited continuously for several centuries, beginning in the Iron Age. Ceramic finds as well as settlement structure

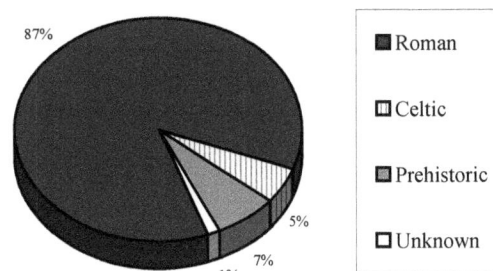

**Figure 8.2**: the chronological distribution of 13,786 identifiable animal bones.

| | Trauma | | Inflammation | | Oral pathology | | Tuberculosis | | Total | |
|---|---|---|---|---|---|---|---|---|---|---|
| | n | % | n | % | N | % | n | % | n | % |
| Cattle | 3 | 5.6 | 8 | 14.9 | 3 | 5.6 | 0 | 0 | 14 | 26 |
| Pig | 1 | 1.8 | 4 | 7.4 | 3 | 5.6 | 0 | 0 | 8 | 14.8 |
| Dog | 4 | 7.4 | 11 | 20.3 | 3 | 5.6 | 14 | 25.9 | 32 | 59.2 |
| Total | 8 | 14.8 | 23 | 42.6 | 9 | 16.8 | 14* | 25.9 | 54 | 100 |

**Table 8.1**: taxonomic distribution of pathological lesions observed in the material. Key: * - affected bones in two individuals.

## Animal diseases observed

In total, 54 pathologically modified bones were found. This may be considered a relatively large number, since the manifestation of pathological phenomena, to a great extent, depends on the richness of archaeozoological assemblages. This may be best demonstrated by comparing traumatic lesions, which are identified most consistently in the literature. Siegel (1976) reported 19 such cases (0.04%) in her review article based on 47,300 excavated animal bones. Among the 54 pathological cases from Balatonlelle-Kenderföld eight cases of traumatic lesions correspond to 0.09% of the NISP recorded at this site.

With the exception of five specimens of uncertain dating, all derive from the Roman period. Naturally, they are far from representative of all animal diseases of the time since many maladies leave no trace on the skeleton. However, using broad, empirical categories, the following deformations were observed in the material: trauma; inflammation; oral pathology; and tuberculosis (Tab. 8.1).

## Traumatic lesions

Traumatic deformations in the skeleton may be caused by a number of agents. These include intra-specific clashes (such as injuries inflicted during the mating season) as well as inter-specific fighting, including hunting by humans. Even peri-mortem trauma may be included in this category. Certain types of disease that weaken osseous tissue may further contribute to these conditions (see Groot this volume). While animal species of smaller body sizes recover from such trauma relatively easily, fractures, especially in the limbs, may lead to death or slaughter in large ungulates (Tamás 1987, 299). In these latter animals, therefore, healed rib fractures seem to have been more common relative to those in long bones.

These general observations are nicely reflected in the material under discussion here: all three traumatic lesions observed in cattle (*Bos taurus* L., 1758) were healed fractures of the rib showed no major dislocation. Conversely, only a single rib fracture occurred in dog (*Canis familiaris* L., 1758), while the other three injuries were found on bones of the extremity in this species. One of these was the fracture of femur in an approximately

one and a half year-old dog, which healed with a grave dislocation, possibly because the musculature had kept the injured limb in a contracted position (Fig. 8.3). This condition may have contributed to the early death of the animal, although the fact that the callus appears to have remodelled is indicative of an advanced healing process. This could indicate that the accident happened at an early stage of the animal's life (R. M. Thomas, pers. comm.). Symptoms of a badly infected compound fracture were observed on the radius and ulna of another individual (Fig. 8.4).

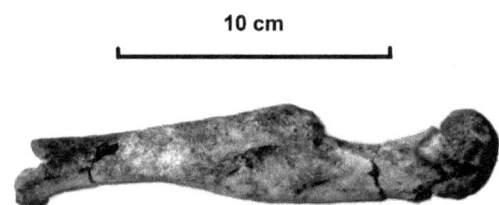

10 cm

**Figure 8.3**: fracture on a Roman dog femur.

10 cm

**Figure 8.4**: fracture on a Roman dog radius and ulna.

Figure 8.5: fracture on a Celtic period pig tibia and astragalus.

It may be presumed that this animal suffered from a severe bone infection, probably caused by trauma inflicted in the shoulder region. A cut mark on this bone may be indicative of the fact that this animal was slaughtered and butchered, possibly in relation to this grave condition. It also suggests that the meat from this animal was not rejected despite the evidently major inflammation. This scapula came to light from a feature that contained mostly ordinary food refuse (bones from caprines, pig, horse, a goose and a hen). It is interesting, but probably coincidental, that remains of one of the tuberculous dogs were also deposited in this feature.

A similarly grave, open fracture has completely distorted the distal tibia and astragalus of a pig (*Sus scrofa* L., 1758; Fig. 8.5). According to Benecke (1994, 166, Fig. 80) this type of injury occurs commonly at archaeological sites in Germany (*e.g.* Haithabu, Heuneburg, Hitzacker, Manching and Ralswiek), and may be caused by tethering the animals by tying a rope on one of their hind legs. These are among the few bones that originate from the Celtic Period, when pig keeping seems to have been of special importance. The relatively great number of pig bones (37%; NISP=538) in the otherwise small Celtic Period component of the faunal assemblage thus offered a greater statistical probability for the manifestation of pathological lesions.

Figure 8.7: inflammation on a Roman period medial phalanx of cattle.

## Inflammations

A syndrome should be understood under the 'catch-all' term of inflammation, rather than a concrete group of diseases. A number of causes may lie behind inflammations, some of which can never be identified. These symptoms occurred randomly in the material, on cervical vertebrae and scapulae, as well as on phalanges. An impressive case was observed on a Roman period cattle scapula. Almost the entire surface of this large flat bone was covered with areas of bone resorption as well as exostoses (Fig. 8.6).

Figure 8.8: inflammation on a Roman period medial phalanx of cattle; proximal view.

Symptoms of another remarkable inflammation were evident on the proximal articulation of a medial cattle phalanx (Figs. 8.7 and 8.8). The formation of such exostoses is usually explained by excess workload in draught animals. Undoubtedly, animals exploited for labour often develop exostoses of this type. Typical symptoms of draught exploitation include not only

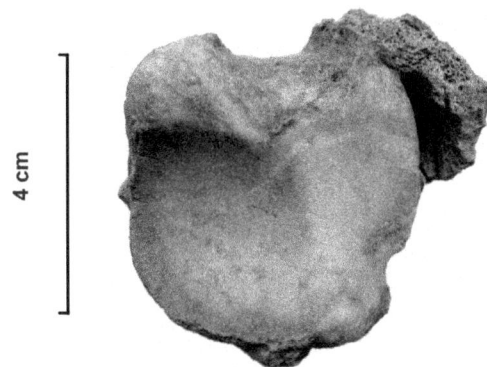

Figure 8.6: exostoses on a Roman period cattle scapula.

osseous growths caused by inflammation, but also exostoses and other deformations in limb bones (Higham *et al.* 1981). Unfortunately, it is usually impossible to tell which of these symptoms originate from generic inflammation and which were caused by excess workload. The latter interpretation is supported by the symmetric location of such exostoses in the limb bones of complete skeletons, as well as the greater degree of inflammation in the bones of the distal extremity (Bartosiewicz *et al.* 1997).

## Oral pathology

Disorders of the teeth are relatively common and easily observed in archaeological materials. This is partly due to the fact that taphonomic agents damage resistant teeth to a lesser extent than other parts of the skeleton. In addition, the oral cavity as the 'entrance' to the digestive system is relatively vulnerable. Naturally, not all disorders of this type should be considered pathological. If, for example, some teeth are unusually heavily worn, one does not necessarily have to consider pathological agents behind this phenomenon. Two cattle teeth showed such heavy wear. Other teeth had heavily inflamed roots causing them to fall out during the animal's life. In one case, the empty alveolus in the mandible of a dog remodelled and closed following the loss of the tooth. Various degrees of this type of pathological deformation were observed in cattle, pig as well as dog in this material. Given the small number of these 10 finds and the fact that percentages may thus be appraised visually, Tab. 8.2 shows only absolute numbers.

|  | Inflammation | Oligodontia and arrested development | Abnormal toothwear | Total |
|---|---|---|---|---|
| Cattle | 1 |  | 2 | 3 |
| Pig | 4 |  |  | 4 |
| Horse |  | 1 |  | 1 |
| Dog | 1 | 1 |  | 2 |
| Total | 6 | 2 | 2 | 10 |

**Table 8.2**: taxonomic distribution of pathological phenomena observed in the oral cavity.

## Tuberculosis

Characteristic osteological symptoms of tuberculosis could be observed in the case of two dogs. This is all the more interesting since tuberculosis infections among domestic animals are most common in poultry and some-times in pig (Baker and Brothwell 1980, 77) and incidences have been rarely reported in the zooarchaeological literature (although see Bendrey, and Csippán and Daróczi-Szabó, this volume). These bacteria may attack any organ in the living animal, although bone deformations emerge only in the later phases of the infection. The skeleton tends to be infected through the

circulatory and lymphatic system sometimes in the proximity of an already infected organ (Mende unpublished). The chief symptom of tuberculosis infection in the skeleton is the gradual but massive erosion of osseous tissue. However, Bathurst and Barta (2004) note that bone cell proliferation rather than demineralisation is a more characteristic response to this disease in dogs.

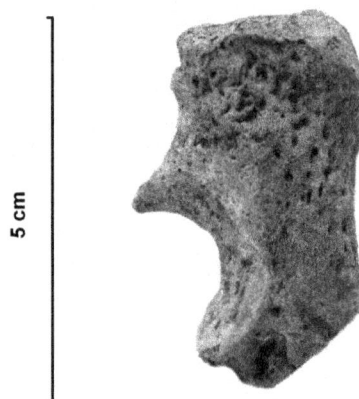

**Figure 8.9**: tuberculosis on a Roman dog ulna.

**Figure 8.10**: tuberculosis on a Roman dog pelvis.

One of the two dog skeletons was almost complete, while the other was represented by only a few bone fragments (the nuchal region of the skull, a pelvis fragment and three ribs). In the first case, almost all bones showed characteristic tuberculous lesions, periosteal proliferation and subsequent erosion (Figs. 8.9 and 8.10). This process completely destroyed the head of the left femur (Fig. 8.11). In the other animal, the most characteristic symptoms could be recognised on the remaining rib fragments in the form of bulbous growths indicative of bone cell proliferation (Fig. 8.12). Another, asymmetrically located outgrowth was found on the pelvis fragment (Fig. 8.13), although it is impossible to tell whether this lesion is related to tuberculosis or results from some form of trauma that caused localised periostitis.

5 cm

Figure 8.11: tuberculosis on a Roman dog femur.

5 cm

Figure 8.12: tuberculosis (bulbous growths) on rib fragments from a Roman dog.

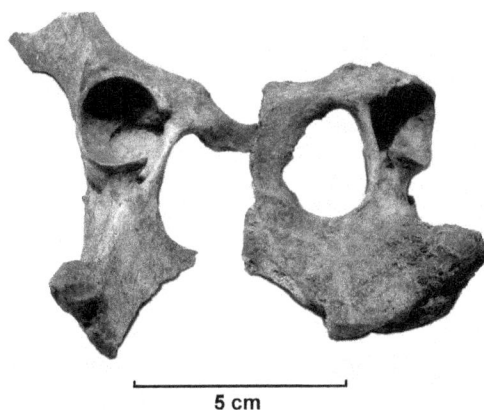

5 cm

Figure 8.13: asymmetrically located outgrowth on a dog pelvis from the Roman period.

Tuberculosis bacteria are easily spread by saliva and excrement *etc.* that contributes to the inter-specific spread of this zoonosis. Dogs may thus have been infected equally by humans or animals. In either case, the presence of tuberculosis may therefore also be hypothesised in the human population.

## Conclusions

The occurrence of pathologically deformed bones in the largest, Roman period component of the material shows the probabilistic nature of these phenomena: presuming random occurrence, rare lesions tend to have a greater statistical chance of being manifested in large bone assemblages (*cf.* Grayson 1985, 136).

Pathological phenomena in cattle included arthritic lesions that may have been caused by draught exploitation. This is to be expected by the Roman period, when these animals were regularly used in tillage. By the Roman period and the Middle Ages an upsurge of such symptoms may be observed in comparison with prehistoric periods (Bartosiewicz 2006).

It is noteworthy that despite the considerable contribution of small ruminants to this assemblage, no pathological phenomena were recorded in this group of animals.

Proportionally speaking, in almost all categories of lesions, a great number of deformations were discovered in dogs, although the contribution of these animals to the entire assemblage was not particularly great. The 14% NISP value calculated for dog, in part, results from articulated skeletal parts. While animals kept for food were often slaughtered before bone deformations could develop, pets and working animals stood not only a better chance of survival, but – in the case of dogs – their complete skeletons or articulated bones make the identification of disease easier.

On the basis of bone remains, at least 28 dogs were represented in the Roman period material. Two of these could be identified as females, while the number of identifiable males was six on the basis of the presence of bacculum (three individuals) and robust cranial structure (five individuals). The 37-65 cm range of withers heights estimated from 52 long bones after Koudelka (1885) shows a variability that may indicate conscious breeding. This is of importance, since the proximity of dogs to humans may have played a part in transferring tuberculosis.

## Acknowledgements

Grateful thanks are due to Prof. Péter Sótonyi for his help with the diagnoses and Dr. László Bartosiewicz who helped with the revision and translation of the manuscript. Constructive comments by Dr. Richard M. Thomas have improved the original manuscript significantly. His contribution is also gratefully acknowledged here.

# Bibliography

Baker, J. and Brothwell, D. 1980. *Animal Diseases in Archaeology.* London: Academic Press.

Bartosiewicz, L., Van Neer, W. and Lentacker, A. 1997. *Draught Cattle: their Osteological Identification and History.* Annalen Zoologische Wetenschappen Vol. 281. Tervuren: Koninklijk Museum voor Midden-Afrika.

Bartosiewicz, L. 2006. Mettre le chariot devant le boeuf. Anomalies ostéologiques liées à l'utilisation des boeuf pour la traction, pp. 259-267, in Pétrequin, P., Arbogast, R.-M., Péterquin, A.-M., Van Willigen, S. and Bailly, M. (eds), *Premiers Chariots, Premiers Araires. La Diffusion de la Traction Animale en Europe Pendant les IVe et IIIe Millénaires Avant Notre Ère.* CRA Monographies 29. Paris: CNRS Editions.

Bathurst, R. R. and Barta, J. L. 2004. Molecular evidence of tuberculosis induced hypertrophic osteopathy in a 16th-century Iroquoian dog. *Journal of Archaeological Science* 31 (7), 917-925.

Benecke, N. 1994. *Der Mensch und Seine Haustiere. Die Geschichte einer jahrtausendealten Beziehung.* Stuttgart: Konrad Theiss Verlag.

Grayson, D. K. 1984. *Quantitative Zooarchaeology.* Studies in Archaeological Science. New York: Academic Press.

Higham, C. F. W., Kijngam, A., Manly, B. F. J. and Moore, S. J. E. 1981. The bovid third phalanx and prehistoric ploughing. *Journal of Archaeological Science* 8 (4), 353-365.

Koudelka, F. 1885. Das Verhältniss der Ossa longa zur Skeletthöhe bei den Säugertieren. *Verhandl. d. Naturforsch. Ver. Brünn* 24, 127-153.

Mende, B. unpublished. *Történeti Antropológia.* Unpublished university coursebook. Budapest: Loránd Eötvös University.

Siegel, J. 1976. Animal palaeopathology: possibilities and problems. *Journal of Archaeological Science* 3 (4), 349-384.

Tamás, L. (ed.) 1987. *Állatorvosi sebészet 2.* Budapest: Mezőgazdasági Kiadó.

# Author's affiliation

Institute of Archaeological Sciences
Loránd Eötvös University
1088 Budapest, Múzeum körút 4/B
Hungary

# 9. Skeletal alterations of animal remains from the early medieval settlement of Bajč, south-west Slovakia

Zora Miklíková

## Abstract

*The aim of this paper is to detail the occurrence of skeletal pathologies from the largest assemblage of animal bones from the early medieval settlement of Bajč, Slovakia. This work represents preliminary results from the doctoral thesis of the author, which is focused on aspects of animal husbandry and hunting during the early medieval period in Slovakia. Altered elements accounted for 0.3% of the entire bone assemblage and were detected exclusively in domestic species. Macro-morphological descriptions and photo documentation of the finds are provided by species. The observed bone changes have been categorised into the following groups of pathological conditions: traumatic injury, disease of joints, oral pathology and other abnormality. Possible cases of infectious disease in horse (Equus caballus L., 1758) and a tumour in pig (Sus scrofa L., 1758) are also presented before the consequences of particular pathological conditions detected in the animals from Bajč are discussed.*

## Introduction

Between 1987 and 1994, a complex archaeological excavation was carried out at the rural medieval settlement in Bajč-Medzi kanálmi (Komárno District, south-western Slovakia; Fig. 9.1C-E). This site, explored by the archaeologists of the Slovak Academy of Sciences, is located on a former peninsula or island surrounded by the Žitava River. The research revealed that the highest density of occupation at the site dates between the sixth and eleventh centuries AD. A total of 551 excavated features have been distinguished by function: houses or semi-subterranean huts, exterior clay ovens, shallow and deep storage pits, roasting pits, further unspecified pits and a system of canals (Ruttkay 2002).

Ongoing analysis of the faunal remains from the site has yielded the largest bone collection from Slovakia dating to the early medieval period. The majority of animal bones belong to domestic mammals, dominated by cattle and caprines. In addition, a wide range of both wild and other domestic species has been identified. In total, 47 complete or almost complete animal skeletons were recovered from pits and houses; these are typical for the eighth to tenth century phase of occupation (Miklíková and Ruttkay 2003). These rare finds and the good state of preservation afford the unique opportunity to explore the health status of particular animals from medieval Bajč and assess how they were used and treated.

The aims of this paper, therefore, are to present the pathological finds from this archaeological site, to provide macro-morphological descriptions of the altered skeletal elements, to suggest their classification according to identified bone changes and to discuss their possible causes and consequences.

## Materials and methods

The samples under study come from the largest assemblage of animal bone remains dating to the medieval period in Slovakia. As mentioned above, a relatively high proportion of the archaeological features contained more or less complete animal skeletons (Fig. 9.1A-B). Thirteen of these complete skeletons exhibited pathologies, presenting a rare opportunity to examine multiple skeletal elements of the same individual. The remainder of the material consisted of fairly fragmented bones representing typical settlement food debris. The different types of preservation mean that some of the finds described in this paper can be considered in their biological and cultural context, while others lack the same level of detail.

The evaluation of the pathological findings was undertaken macroscopically, although some of the bones needed to be studied radiographically, and molecular-genetic techniques are currently being employed to investigate one specimen. Each sample was photographed and stored at the Institute of Archaeology at the Slovak Academy of Sciences.

## Results

Among 12,067 analysed specimens 40 skeletal alterations were identified. This number represents 0.3% of the entire bone assemblage. The complete list of the detected changes together with their biological and cultural context is given in the Appendix.

The observed bone changes can be categorised after Baker and Brothwell (1980) into the following groups of pathological conditions: traumatic injury (TI); disease of

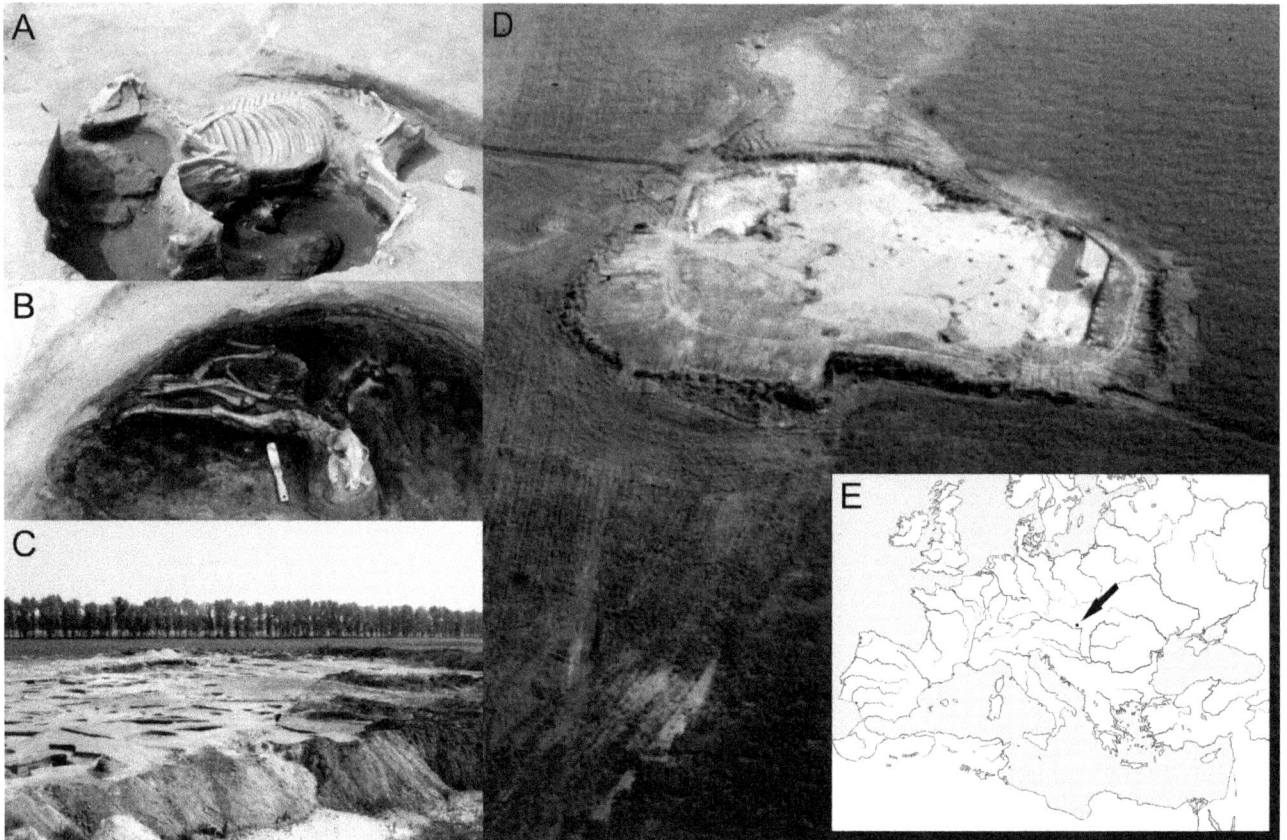

**Figure 9.1**: A - Horse grave (feature 65A); B - Remains of dog deposited in a waste pit (feature 51); C - View of part of the excavation; D – Aerial view of the site; E - Location of the archaeological site in Europe (© Institute of Archaeology of Slovak Academy of Sciences).

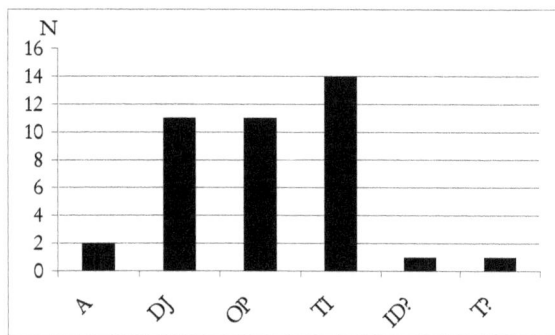

**Figure 9.2**: pathological conditions observed among early medieval animals from Bajč (N = number of fragments).

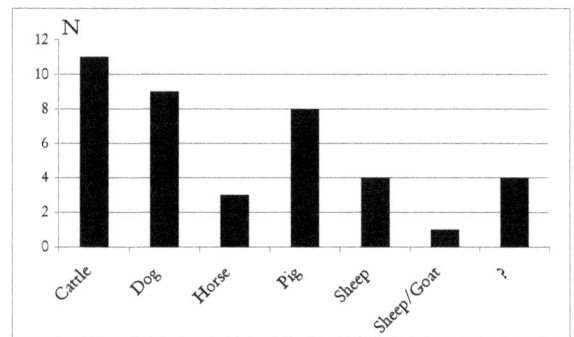

**Figure 9.3**: species distribution of observed pathological conditions (N = number of fragments).

joints (DJ); oral pathology (OP); and abnormality of skeletal development (A). Possible cases of infectious disease (ID) and tumour (T) have also been identified. Bone changes caused by traumatic injury were most abundant, although diseases of joints and oral pathologies were also quite common (Fig. 9.2).

Bone pathologies were found on the following domestic species: cattle (*Bos taurus* L., 1758), horse (*Equus caballus* L., 1758), sheep/goat, pig (*Sus scrofa* L., 1758) and dog (*Canis familiaris* L., 1758). However, it is worth observing that the altered skeletal elements of dog come exclusively from more-or-less complete skeletons of buried animals. No bone alterations of wild species were identified within this archaeofaunal assemblage.

The species distribution of the pathological conditions is presented in Fig. 9.3. Cattle, dog and pig were the most frequently affected domestic species. In four cases it was not possible to conclude species determination due to the incompleteness of the examined bone.

## Cattle

Detected bone changes in cattle (N=11) were basically of three types: joint disease; oral pathology; and developmental abnormality. The most frequently affected skeletal elements were the teeth (Fig. 9.4), lumbar portion of the vertebral column and parts of the appendicular skeleton such as autopodium (tarsal bones) and acropodium (phalanges).

In addition, one horn-core was anomalous in shape. The core belonged to small-horned cattle with a basal circumference of 130 mm and, according to criteria given by Armitage and Clutton-Brock (1976), to a male animal. The horn-core is short in proportion to the basal circumference and rather flattened and oval in cross-section. A striking groove or depression runs along the whole horn-core length, located anteriorly and inferiorly on the body of *processus cornualis* (Fig. 9.5). The explanation for this phenomenon is rather difficult, but it most probably reflects the anomalous development of the horn itself and was not caused by the use of this animal for traction or other work. A bovine scapula with a cross-shaped lesion in the middle of the articular surface, possibly caused by a developmental defect during ossification, was also recorded (Fig. 9.6).

In the area of a house (feature 715) dated to the eighth and ninth centuries AD, a pit of unknown function was excavated. Among other animal bone refuse, part of a cattle backbone was deposited. This specimen comprised six well-preserved lumbar vertebrae and the cranial part of the sacrum, which belonged to one individual older than eight years (Fig. 9.7A). The vertebrae were considerably larger in size compared with the other cattle remains from the site, so it is highly probable that the animal was a castrated male. The pathological alterations appeared on the distal part of the preserved spine – the last lumbar and first sacral vertebrae (L5-S1). The most pronounced bone changes were observed in the sixth lumbar and first sacral vertebrae (Fig. 9.7B). Their articular surfaces are surrounded by distinct osteophytes and the edges of the processus *articularis caudalis* are widened, having developed a lip-like structure. Pitting and numerous perforations could be observed on the caudal surface of the body of the sixth lumbar vertebra and the cranial articular surface of the *basis ossis sacri*. Areas of dense, highly polished bone – eburnation – are also pre-sent (Fig. 9.7C). In the case of L5 and L6 the processus *articularis caudalis* have fused with the *processi mamillo-articularis* (see arrow in Fig. 9.7A).

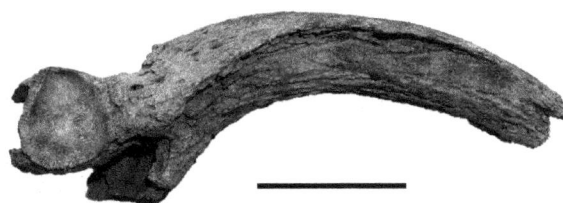

Figure 9.5: cattle horn-core with distinctly deep groove/depression located anteriorly and inferiorly. Scale bar: 50 mm.

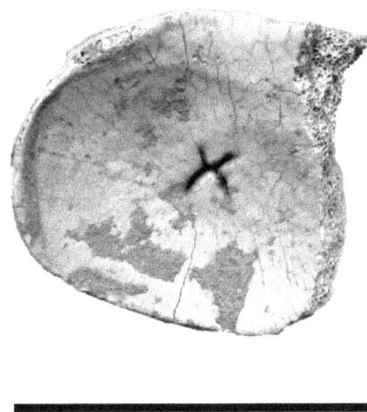

Figure 9.6: a bovine scapula with a cross-shaped developmental defect in the articular surface. Scale bar: 50 mm.

The pathological lesions described above are associated with the condition called *spondylarthrosis* or *spondylosis deformans et ankylopoetica* (*e.g.* Baker and Brothwell 1980; Šutta *et al.* 1986; Zendulka *et al.* 1987). According to Baker and Brothwell (1980, 129-133) *spondylosis deformans* (vertebral osteophytosis) is relatively common in archaeological material. The pathogenesis of this condition is not entirely clear, but in this case the primary lesion was probably focal degeneration of the annulus fibrosus leading to the formation of intra-discal fissures. This would have resulted in the rupture of the annulus and loss of disc tissue, causing the exposed bone at the ends of the vertebral bodies to become ground and polished. The degeneration of the annulus would have al-so lead to the breakdown of the outermost layer of fibre – mainly ventrally and laterally – which would have become necrotic. Ossification of newly formed fibro-cartilage would have lead to the formation of osteophytes, which in their most advanced form would have lead to the union of vertebrae. The causal factors behind such bone changes are variable and may be a combination of age, inflammation, malnutrition, trauma, or overloading. On the basis of a very similar archaeofaunal find of bone pathology from Svodín, also located in south-west Slovakia, it has been suggested that such lesions could have developed as a result of draught use (Fabiš 2005). This castrated male was found in one of the cattle burials of the Baden culture settlement of the site (Eneolithic period – second half of the fourth millennium BC). It seems plausible to suggest that identical pathological

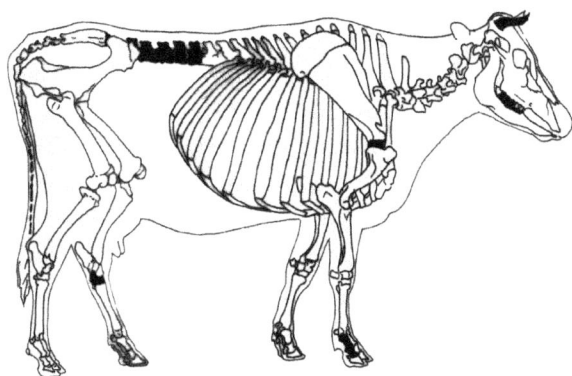

Figure 9.4: skeletal distribution of pathological cattle bones from Bajč.

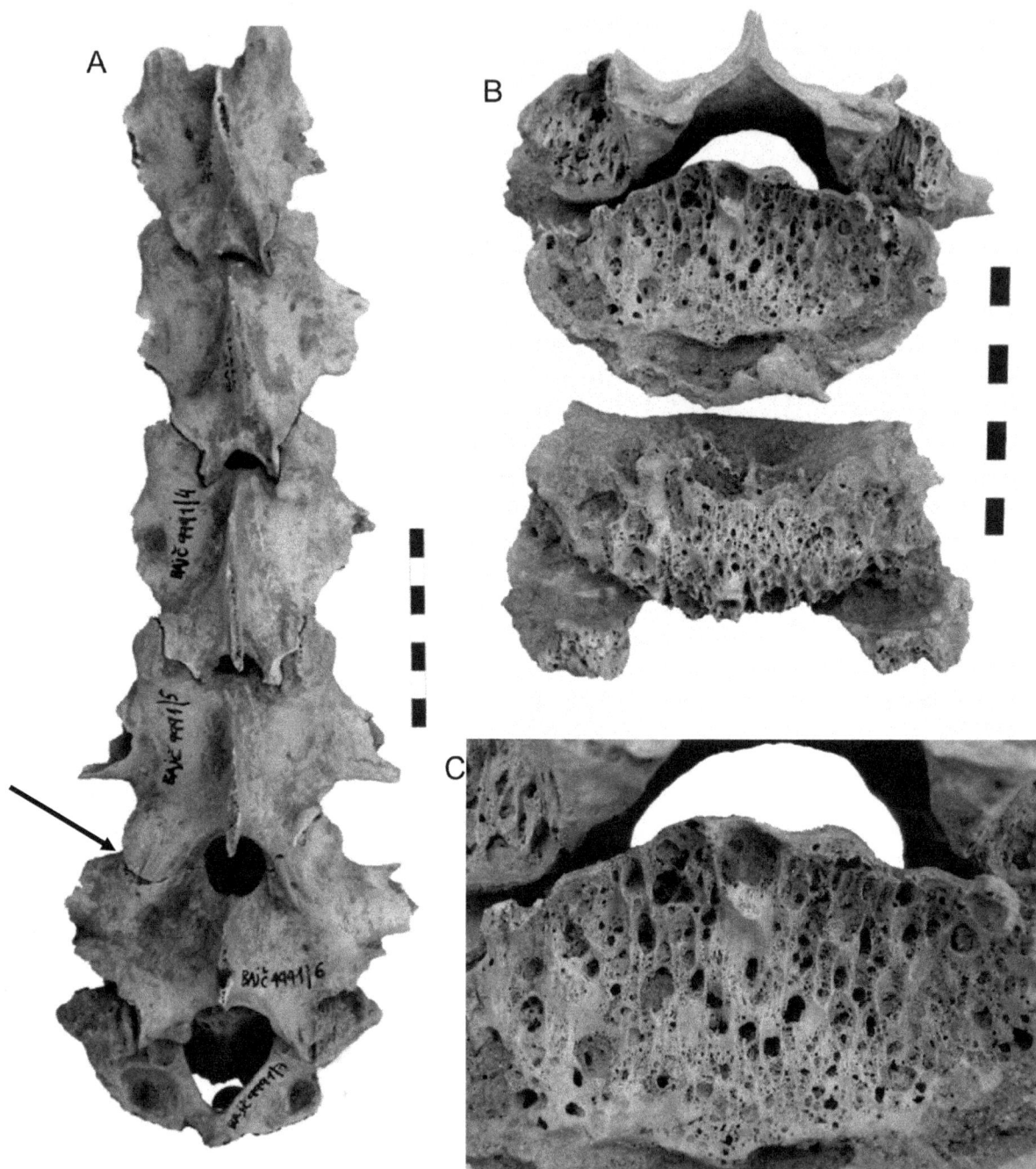

**Figure 9.7**: A - Preserved lumbo-sacral part of cattle backbone; B - affected sixth lumbar (above) and first sacral (below) vertebrae; C - detailed view of the articular surface with highly distinctive bone changes. Scale bar: 50 mm.

identical pathological changes occurring on the remains of the ox from Bajč developed under the same life conditions, and that the animal was used in traction. Unfortunately, no other parts of the ox skeleton were recovered from the pit and we could not say whether there were other bone changes related to this condition.

The fused centro-tarsal and second and third tarsal bone from left cattle leg may represent a case of spavin (Fig. 9.8). According to Baker and Brothwell (1980, 117), "spavin is principally a disease of the tarsus of the horse although it has occurred in trek oxen, draught cattle and camels. Typically it affects the small bones of the inner lower aspect of the joint, causing exostoses, which limits its movement". These authors (1980, 118)

distinguish a number of different types of this condition, including occult spavin, which does not produce any significant exostoses on the small tarsals. This condition is very similar to the specimen from Bajč, where exostoses are absent. Unfortunately, other parts of the appendicular skeleton were not found within the pit material. Bartosiewicz *et al.* (1997) described similar pathological changes in Romanian cattle used as draught animals and recognised more fully developed stages of the disease. They observed that fusion in the centro-tarsal region of the leg represents the advanced case of the disease and usually does not occur earlier than eight years of age. Did the Bajč animal suffer from this kind of pathological condition? If so, it probably did not influence the quality of life of the individual greatly. Baker and Brothwell (1980, 119) suggest that this kind of bone change generally causes only mild degree of lameness and "in many cases, with time and rest, if the condition is not too extensive, the joint will ankylose and the animal become useful again for slow work".

**Figure 9.8**: fused left centro-tarsal with second and third tarsal bones in cattle; anterior view. Scale bar: 50 mm.

Other skeletal alterations might also serve as evidence for the draught use of cattle. Different types of pathological deformations (for example, lipping of the articular surface and exostoses near the proximal or distal articular surface; *i.e.* osteoarthritis) were recognised by Bartosiewicz *et al.* (1997) as possible consequences of using cattle for traction. In their work they suggested the classification of these conditions in phalanges and other affected parts of the appendicular skeleton. While analysing cattle bone material from Bajč similar pathological changes were recorded, although they rarely reached an advanced stage. Both anterior phalanges showed proliferative (exostoses) and degenerative (extension of joint surfaces) changes resulting in malformation of the original bone.

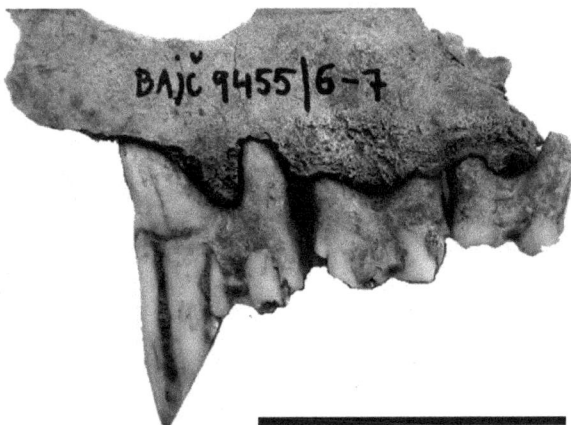

**Figure 9.9**: irregular tooth wear in cattle maxilla. Scale bar: 50 mm.

With regard to oral pathology in cattle, a distinctive example of irregular tooth wear in a fragment of maxilla was observed (Fig. 9). The aboral part of the left upper third molar projects significantly above the normal occlusal line. This could have been caused by several factors, such as the congenital absence or premature loss of the opposite teeth, trauma or partial destruction (Baker and Brothwell 1980).

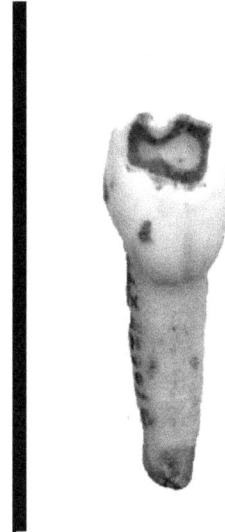

**Figure 9.10**: hypoplastic enamel defect of second incisor.

In addition, a slightly irregular occlusal line was identified in the bovine mandible of an old individual, while a hypoplastic defect of enamel on a cattle incisor is probably a developmental abnormality (Fig. 9.10). Finally, the second lower premolar in one mandible was turned in its vertical axis and resulted in the slight crowding of the premolars of this individual. Such a condition is more common in pigs and dogs and is considered to be a consequence of domestication.

## Dog

Traumatic injuries were frequently noted amongst the dog bones from Bajč, including numerous finds of healed fractures as evidenced by callus formation, foreshortening and displacement of individual elements. These lesions were typically located on the trunk, skull and extremities of the complete and partial dog skeletons (Fig. 9.11). Four out of the five adult dog skeletons from Bajč exhibited fractures of ribs and tibiae in various stages of healing.

One possible dental fracture was also recorded; a mandibular canine that had been split ante-mortem, as revealed by the remodelling of the fractured edges of the tooth.

One peculiar find that has been diagnosed by veterinarian and archaeozoologist Marián Fabiš was three dog metacarpals from the same individual showing changes connected with pseudoarthrosis and *periostitis chronica*, possibly resulting from an infected traumatic injury (Fig. 9.12). In the same adult dog, further traumatic

67

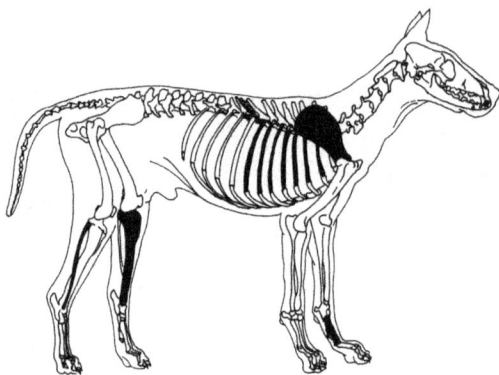

Figure 9.11: the distribution of traumatic injuries in dog.

insults were apparent, including a small opening in the scapula blade accompanied by local periostitis and small exostoses on the corpus of *os ischii*. Osteoarthritic deformations on the lumbar and thoracic portions of the vertebral column identified in another dog skeleton (Fig. 9.1B) are most likely the result of old age (Fig. 9.13). Most of the described injuries probably represent consequences of sudden trauma caused by humans or other animals and emphasise the violent environment of that time, both cultural and natural. The morphology of dogs from the site suggests that they belong to a herding or guarding breed, so their contacts with both humans and animal predators belong to day-to-day activities. Consequently, they were more likely to have been frequently exposed to traumatic events.

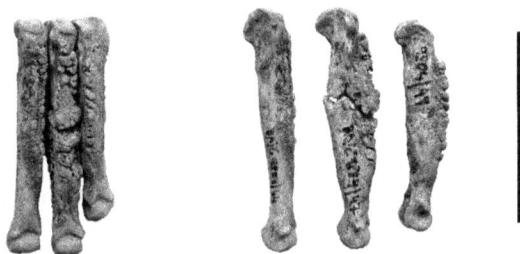

Figure 9.12: possible pseudoarthrosis and *periostitis chronica* in dog metacarpals. Note the incompletely healed fracture in the middle specimen on the right; anterior view (left) and lateral view (right). Scale bar: 50 mm.

## Horse

Among the horse bone remains from Bajč, three pathological lesions were detected. Ascertaining the cause of these changes is rather problematic; consequently, their classification in this article is only preliminary. The pathological condition known as periostitis is evident on a horse metapodial. This inflammatory process results in the proliferation of new bone, which, in this specimen, was deposited on the outer compact bone surface of the diaphysis. The deposition of new bone tissue appears localised to the middle of the shaft fragment, although an even larger area of the proximal part of the metapodial has an unusually porous

Figure 9.13: thoracic vertebrae with osteoarthritic bone changes in the *fovea costalis*; lateral view. Scale bar: 50 mm.

surface (Fig. 9.14). From inside the diaphysis a circular depression (fistula?) had been created.

Similar changes were recorded in two sternal ribs originating from a horse buried in a single grave (Fig. 9.1A). These were plaque-like depositions of periosteal bone concentrating on the visceral surface of the rib (Fig. 9.15). It has been suggested that this kind of periostitis strongly resembles symptoms of infectious disease similar to tuberculosis in humans (Lenka Vargová and Ladislava Horáčková pers. comm.). In order to verify this hypothesis, the sample has been submitted for molecular-genetic analysis to determine if the *Mycobacterium tuberculosis* bacteria could be detected. This male horse died as a sub-adult (*i.e.* 2.5-3.5 years old) and there were no other related bone changes detected on the skeleton. It is possible that the as-yet unspecified disease from which this horse suffered was the cause of its early death. At the very the least recorded bone changes indicates that the animal suffered from inflammation or infection of the pleura before its death.

The bone exostoses on the first phalanx posterior of older horse individual may have developed as a result of prolonged use, although age, sex, weight and foot conformation may have also contributed to the osteophytic growth.

## Pig

The creation of local, ossifying periostitis was recorded on one pig tibia. This individual probably suffered from a sudden or chronic insult to the bone. One pig humerus also shows the features of a tumour or tumour-like cyst (Fig. 9.16).

The remains of trauma in pigs were rare; only one metacarpal bone showed a healed fracture of the diaphysis. However, oral pathologies were very frequent amongst the pig remains. Three cases of rotated premolars were recorded – two maxillary first premolars and one mandibular fourth premolar were turned in different axes. A strongly abraded upper third molar with atypical wear of the proximal part of its roots (from the

68

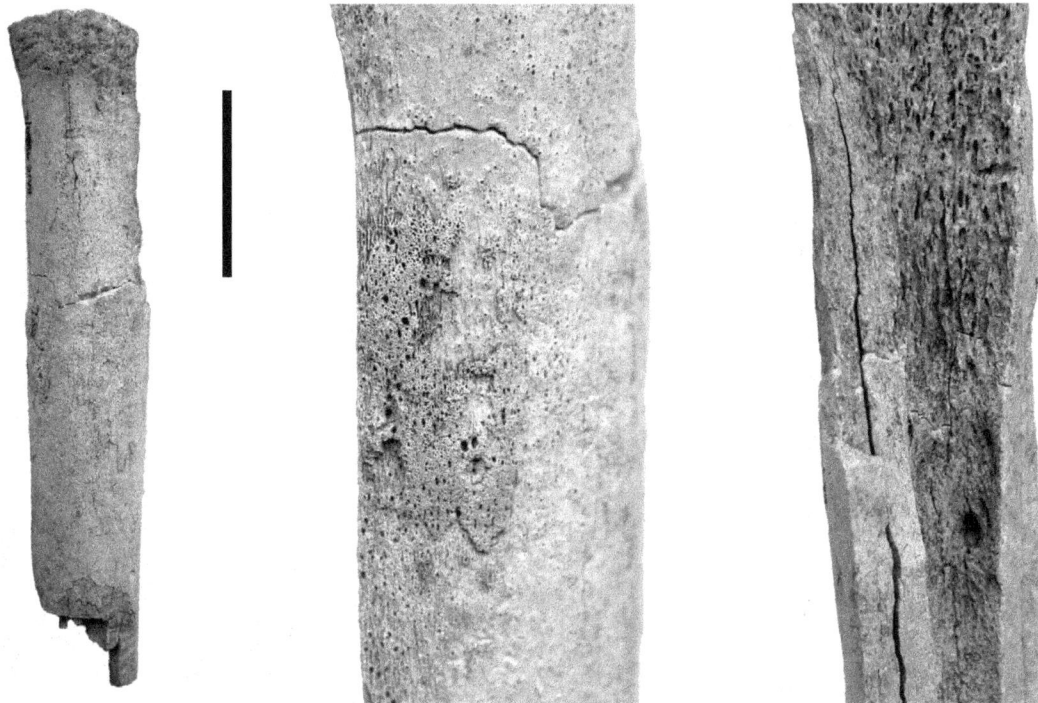

**Figure 9.14**: a fragment of horse metapodial affected by periostitis (left: outer surface, middle: detail on affected proximal shaft area, right: internal surface). Scale bar: 50 mm.

**Figure 9.15**: a plaque-like bone deposition on a horse rib – a possible case of infectious disease.

buccal side) was also noted, although the cause of this defect is unknown (Fig. 9.17). A periodontal abscess was detected in a well-preserved mandible of an adult female. This was located at the apex of the roots of both right premolars and probably resulted in the ante-mortem loss of the second and fourth premolars. The reason for this condition could be pulp exposure, caries, calculus or periodontal disease (Baker and Brothwell 1980).

Figure 9.16: a tumour-like cyst on the diaphysis of the distal humerus of a pig. Scale bar: 50 mm.

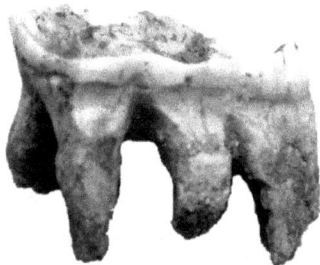

Figure 9.17: pig third molar with anomalous groove located below the crown; buccal view. Scale bar: 50 mm.

## Sheep and Goat

Advanced cases of osteoarthritis were found in a sheep skeleton uncovered from a small pit located in the middle of the house area (Fig. 9.18). The house was dated to the ninth century AD and the adult and foetal sheep are considered to have been sacrificed during the consecration of the building. The remains of the older individual consisted of the skull and foot bones. This animal was most likely chosen to be sacrificed on the basis of a handicap discovered during analysis. The right metatarsal of this animal showed extension of the distal articular surface and of the adjacent articular surfaces of the proximal phalanges (Fig. 9.19). Eburnation of the distal articular surface of these bones was present together with massive osteophytes. Finally, the second and third phalanges had fused, which would have resulted in lameness. Other examples of joint disease were recorded in a sheep talus (eburnation of the articular surface of caput tali) and the first phalanx of an adult individual (exostoses). Oral pathology was represented by two cases. In one complete mandible of an adult sheep, alveolar bone recession is visible and in the second case, the crowding of mandibular teeth of a juvenile sheep was detected.

## Conclusions

On the basis of the palaeopathological findings, recorded during the archaeozoological study of the early medieval rural settlement in Bajč, an attempt to study the health status of the animals living within the site was made. In total, skeletal alterations affected 0.3% of the assemblage. These results concord with the recognised rarity of palaeopathological cases; they usually only constitute 1-1.5% of entire archaeofaunal assemblage (Fabiš, pers. comm.). From Slovakia, information on the prevalence of pathologies in archaeofauna in relation to period and region is rare. Fabiš (2004) has recorded the 1.4% occurrence of pathological finds within one of the small archaeofaunal assemblages (NISP=361) from south - western Slovakia, dated to the Iron Age period. For our region only future research will bring more exact results for potential comparative studies (e.g. Vondráková et al. 2007).

In Bajč, we can speak about three main types of pathological conditions, from which some of the animal individuals suffered: traumatic injuries, diseases of joints and oral pathologies. Cases of skeletal anomalies, possible cases of tumour and infectious disease, were also described. Some recorded bone abnormalities, probably of congenital origin, are perhaps unlikely to have affected the animals. In contrast, cases of probable infectious disease in horse or joint disease resulting in lameness might have been the reason for the early death of at least one animal.

Most of the pathological conditions exhibited amongst the Bajč material have their parallels in the palaeopathological literature, but among the documented bone assemblages from Slovakia, some of them show unique aspects in regard to their manifestation. The frequency of certain pathological bone changes and the specific species that were affected, possibly reflect environmental and cultural conditions under which the animals were kept. The skeletal alterations discussed above show that cattle and dog were the two most frequently affected species at Bajč: cattle suffered mostly from work-load disease or stress and dogs from traumatic injuries. Different attitudes to various animals suffering from disease seem highly possible depending on their economic value and social status. This has been shown after considering the described finds in their archaeological context: for example, while the horse affected by infection had been buried and attempts may have been made to cure the animal, the sheep that suffered from osteoarthritis was probably sacrificed on the basis of its handicap.

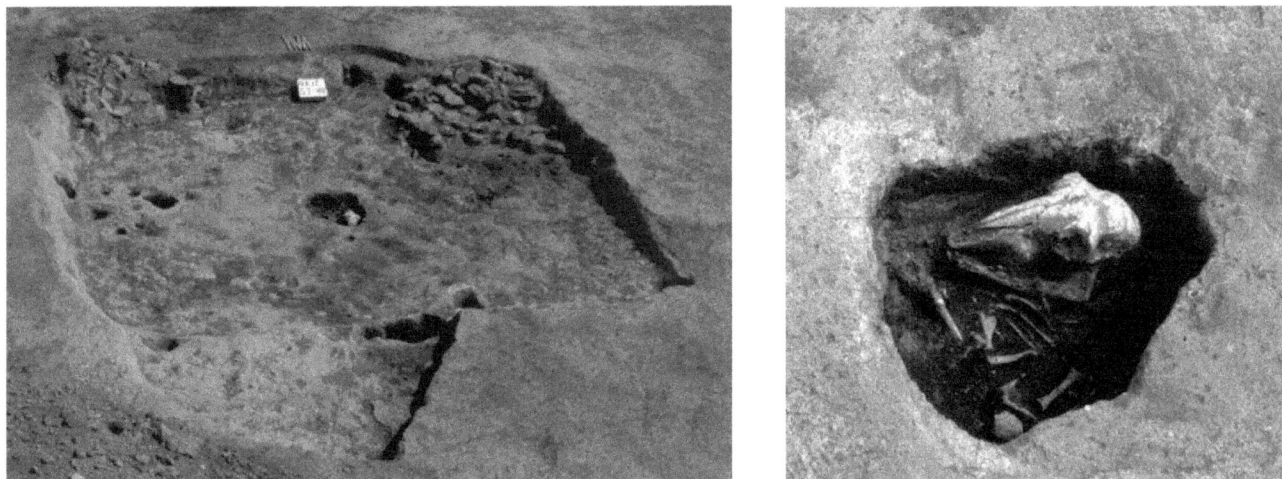

Figure 9.18: a house area (left) and a pit with the adult and foetal/neonatal sheep remains (right; photo by M. Ruttkay).

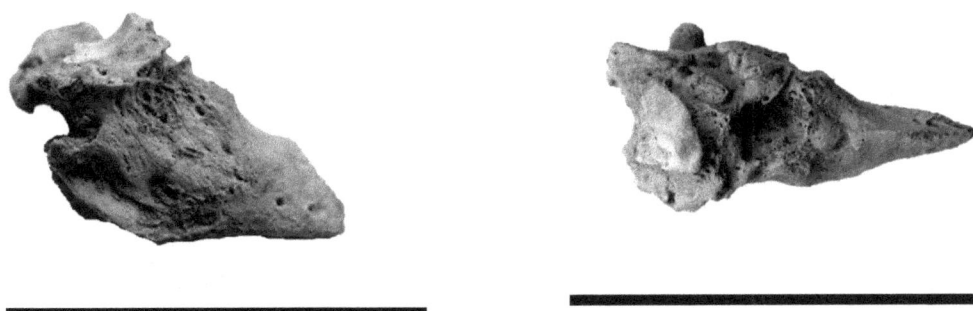

Figure 9.19: advanced case of osteoarthritis in sheep foot bones – fused second and third phalanges (left: lateral view, right: dorsal view). Scale bar: 50 mm.

## Acknowledgements

I am grateful to MVDr. Marián Fabiš PhD who supervised me during my doctoral study focused on archaeozoology at the Institute of Archaeology, for all his sup-port and motivation. I am also thankful for his diagnosis of some of the bone pathologies presented in this article.

The animal bone assemblage and field photographs used in this analysis were made available by Dr. Matej Ruttkay, field director of the excavation in Bajč (Institute of Archaeology of Slovak Academy of Sciences).

Special thanks go to Laura Niven, PhD for editing my English and to Richard Thomas, PhD for his valuable comments on this text.

## Bibliography

Armitage, P. L. and Clutton-Brock, J. 1976. A system for classification and description of the horn-cores of cattle from archaeological sites. *Journal of Archaeological Science* 3, 329-348.

Baker, J. R. and Brothwell, D. R. 1980. *Animal Diseases in Archaeology*. London: Academic Press.

Bartosiewicz, L., Van Neer, W. and Lentacker, A. 1997. *Draught Cattle: their Osteological Identification and History*. Annals of Scientific Zoology, Vol. 281. Tervuren: Royal Museum of Central Africa.

Fabiš, M. 2004. Palaeopathology of findings among archaeofaunal remains of Small Seminar site in Nitra. *Acta Veterinaria Brno* 73, 55-58.

Fabiš, M. 2005. Pathological alterations of cattle skeletons - evidence for the draught exploitation of animals? pp. 58-62, in Davies, J., Fabiš, M., Mainland, I., Richards, M. and Thomas, R. (eds), *Diet and Health in Past Animal Populations: Current Research and Future Directions*. Oxford: Oxbow.

Miklíková, Z. and Ruttkay, M. 2003. Archaeozoological analysis of animal deposits from the medieval settlement in Bajč, pp. 207-216, in Hašek, V., Nekuda, R. and Unger, J. (eds), *Service to Archaeology IV. Proceedings from the 5th Work Conference Scientific Methods in Archaeology and Anthropology held in 22nd-24th May 2002 at Masaryk University in Brno.* Nitra: Muzejní a vlastivědná společnost v Brně, Geodrill Brno, AÚ SAV Nitra.

Ruttkay, R. 2002. Mittelalterliche Siedlung und Gräberfeld in Bajč-Medzi kanálmi (Vorbericht). *Slovenská Archeológia* 2, 245-322.

Šutta, J., Orságh, A., Janda, J., Kottman, J., Král, E., Nechvátal, M. and Roztočil, V. 1986. *Veterinárna chirurgia.* Bratislava: Príroda.

Vondráková, M. Matejovičová, B., Vondrák, D., Kolena B., Ambros, C., Fabiš, M., Miklíková, Z., Martiniaková M., Bauerová, M., Bauer, M., Omelka, R. 2007: *Katalóg paleopatologických nálezov z archeologických výskumov na Slovensku.* FPV UKF v Nitre, Edícia Prírodovedec č. 290, (CD).

Zendulka, M., Škarda, R., Černý, L., Groch, L., Halouzka, R., Holman, J., Kaman, J., Konrád, V., Marcaník, J., Pauer, T. and Pivník, L. 1987. *Patologická anatomie hospodářských zvířat.* Praha: SZN.

## Author's affiliation

Institute of Archaeology
Slovak Academy of Sciences
Akademická 2
949 21 Nitra
Slovakia

# Appendix.

List of detected skeletal alterations from Bajč (L – large mammal, M – medium mammal, L/M – large or medium mammal).

| Species | Skeletal element | Side | Sex | Age | Bone changes | Category | Invent. Nr. | AD | Feature |
|---|---|---|---|---|---|---|---|---|---|
| Cattle | scapula | R | ? | subadult/adult | cross-shaped lesion in the middle of the distal articular surface | A | 9953/13 | 8-9 | 700 |
| Cattle | os cornu | L | F? | subadult/adult | longitudinal groove on the body of horn-core | A | 9142/14 | 8-9 | 816 |
| Cattle | phalanx I, anterior | ? | ? | subadult/adult | exostoses near the distal end, widening of the proximal articular surface | DJ | 9717/13 | 9-11 | 18 |
| Cattle | vertebra lumbares | R+L | ? | subadult/adult | exostoses | DJ | 9777/4 | 9 | 54 |
| Cattle | os centrotarsale, os tarsale II, III | L | ? | subadult/adult | fusion of os centrotarsale and os tarsale II+III, porous surface | DJ | 9302/5 | 8-9 | 925 |
| Cattle | phalanx II, anterior | ? | ? | adult | lipping at the proximal articular surface, exostosis near the distal end | DJ | 9598/4 | 9 | 1067 |
| Cattle | vertebra lumbares, sacrales | R+L | C? | subadult/adult | exostoses, lipping, highly polished bone | DJ | 9991/1 | 8-9 | 715B |
| Cattle | dens inferior | R | ? | adult | rotation of P1 | OP | 9012/11-12 | 8-9 | 729 |
| Cattle | dens inferior | R | ? | adult | atypical tooth wear | OP | 9014/1-2 | 7-8 | 730 |
| Cattle | dens inferior | L | ? | subadult/adult | enamel hypoplasia on buccal surface of I1 | OP | 9238/6 | 8-9 | 886 |
| Cattle | dens superior | R | ? | adult | atypical tooth wear | OP | 9455/6-7 | 8-9 | 1000 |
| Dog | vertebra thoracales, lumbares | R+L | M | 7 years+ | arthrosis | DJ | 9772/47 | 8 | 51 |
| Dog | pelvis | R | F? | adult | small exostoses above acetabulum | DJ | 9804/24 | 8 | 319 |
| Dog | fibula | R | M | 7 years+ | fracture healed with callus | TI | 9772/49 | 8 | 51 |
| Dog | dens inferior | R+L | M | adult | broken canine, polished edges | TI | 9787/74-75 | 9-10 | 88 |
| Dog | costae | ? | ? | adult | fracture healed with callus | TI | 9789/3 | 9 | 93 |
| Dog | scapula | R | F? | adult | inflammatory process around small opening in the thoracic part of blade | TI | 9804/8 | 8 | 319 |
| Dog | metacarpus III,IV,V | L | F? | adult | pseudoarthrosis, chronic periostitis, exostoses | TI | 9804/44,47,49 | 8 | 319 |
| Dog | tibia | L | M | adult | fracture healed with callus | TI | 9807/15 | 8 | 904 |
| Dog | costea | ? | M | adult | fracture healed with callus | TI | 9807/31 | 8 | 904 |
| Horse | phalanx I, posterior | L | M | 12-13 years | exostoses near the distal end, medially | DJ | 9718/19 | 9-11 | 18 |
| Horse | costae | L | M | 2.5-3.5 years | periostitis | ID | 9873/9869/111 | 8-9 | 65A |
| Horse | metatarsus III | ? | ? | subadult/adult | inflammatory process, creation of new bone tissue | TI | 9860/2 | 9-11 | 79A |
| Pig | humerus | R | ? | subadult/adult | tumour-like cyst? | N | 9432/12 | 8-9 | 983 |
| Pig | dens superior | L | ? | adult | atypical tooth wear of M3 on the lingual surface | OP | 9978/4 | 7-8 | 714 |
| Pig | maxilla | R | F | subadult/adult | rotation of P1 | OP | 9012/9-10 | 8-9 | 729 |
| Pig | dens inferior | L | F | subadult/adult | abscess beneath P2 | OP | 9050/1-3 | 8-9 | 750 |
| Pig | dens inferior | R | ? | adult | rotation of P4 | OP | 9317/20 | 7-8 | 931 |
| Pig | dens superior | L | ? | subadult/adult | rotation of P1 | OP | 9435/5-6 | 8-9 | 983 |
| Pig | tibia | R | ? | subadult/adult | creation of new bone tissue on the diaphyseal surface | TI | 9170/10 | 8-9 | 839 |
| Pig | metacarpus V | L | ? | juvenile/subadult | fracture healed with callus | TI | 9594/4 | 8-9 | 1065 |
| Sheep | phalanx I,II, III, posterior | R | F? | adult | exostoses and lipping of both phalanges I, fusion of phalanges II and III | DJ | 9776/57-59 | 9 | 53 |
| Sheep | phalanx I, anterior | ? | ? | subadult/adult | exostoses near the distal end | DJ | 9801/60 | 8-9 | 304 |
| Sheep | talus | L | | subadult/adult | arthrosis, eburnation of the caput tali | DJ | 9206/5 | 8-9 | 867 |
| Sheep | dens inferior | R+L | F? | adult | atypical tooth wear P4 and M1, paradentosis | OP | 9776/35 | 9 | 53 |
| Sheep/goat | dens inferior | R | ? | infant/juvenile | crowded teeth | OP | 9317/8-9 | 7-8 | 931 |
| L | cartilago costae | ? | ? | subadult/adult | fracture healed with callus | TI | 9369/9 | 8-9 | 948 |
| L/M | costae | ? | ? | subadult/adult | fracture healed with callus | TI | 9181/9 | 8-9 | 846 |
| M | costae | ? | ? | subadult/adult | fracture healed with callus | TI | 9892/30 | 9 | 327 |
| M | costae | ? | ? | subadult/adult | fracture healed with callus | TI | 9921/6 | 8-9 | 344 |

73

# 10. Animal diseases from medieval Buda

Péter Csippán & László Daróczi-Szabó

## Abstract

*According to written documents, the castle of Buda was established on the Castle Hill by Béla IV in the middle of the thirteenth century. During the centuries, Buda gradually developed to become the political, commercial and military centre of medieval Hungary. Archaeological research has been going on here since World War II. One of the most important excavations took place on the south-western part of today's St. George Square. This excavation covers every period of the history of Buda, from the prehistoric age until the twentieth century. Thousands of animal bones were recovered in this excavation; however, signs of disease and trauma were only rarely observed. This may be due to the culling of sick animals; livestock may have been largely killed before the symptoms of diseases became manifest in skeletal tissue. The pathological symptoms identified occurred more commonly in pets (cats and dogs etc.), probably because these animals had better chances of treatment and survival into old age. In this paper, we have chosen to present pathological bone finds from three important periods of Buda: the end of the Árpád Period (later thirteenth century); the Sigismund Period (fourteenth-fifteenth century); and the Period of the Ottoman Turkish occupation (sixteenth-seventeenth century). Our paper presents some pictures of healed fractures with dislocation, signs of tuberculosis, osteoarthritis, healed fractures with false joint formation and joint inflammation etc. These diseases are individual cases, but sometimes they may be considered characteristic on different parts of the skeleton in a variety of animal species.*

## Introduction

Documentary evidence reveals that Buda Castle was established by King Béla IV in the middle of the thirteenth century. As the settlement grew and its population density increased, it became the political, commercial and military centre of Hungary.

From the twentieth century, several excavations have been carried out in the area of the Castle. Large scale explorations in the south-western corner of St. George Square began in 1995. This large area is important from an archaeological point of view for several reasons. The first Jewish quarter of medieval Buda stood here between the establishment of the Castle and the 1360s (Zolnay 1987, 28). However, aside from scarce written references, little is known of this community. After 1360, when King Louis I drove out Jewish inhabitants from the area, wealthy Christian families moved in (Magyar 2003, 50).

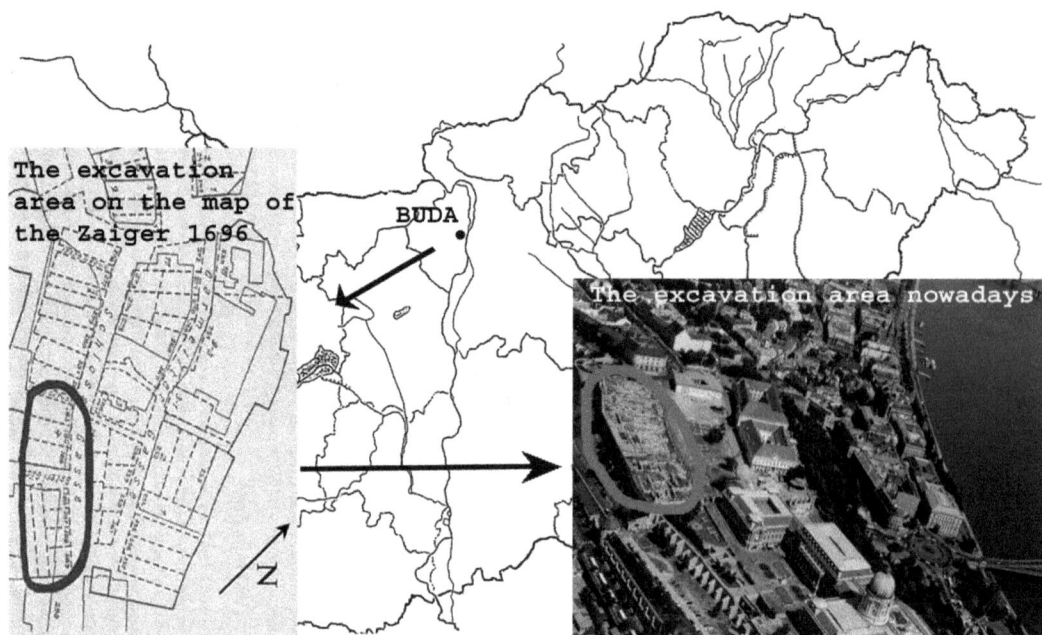

Figure 10.1: map showing the location of St. George Square, Budapest, Hungary.

**Figure 10.2**: symptoms of tuberculosis on the vertebrae and ribs of a dog (period of the Árpád Dynasty).

assemblages have been published from the late thirteenth and early fourteenth century in Hungary in general and Budapest in particular. Only two pathological animals were recorded from this collection.

On several vertebrae and ribs of a dog *(Canis familiaris* L., 1758), symptoms of tuberculosis were observed (Fig. 10.2). The degenerative deformation of the vertebra may be regarded as a typical symptom of tuberculosis. The amorphous tubers that developed on the rib indicate tuberculosis as well. As the animal lived in the town, it is likely that it was infected by human tuberculosis *(Mycobacterium tuberculosis)* The *M. tuberculosis* bacterium usually causes a general infection in carnivores that spreads to the bones (Varga *et al.* 1999, 25).

The other case of pathological deformation from the Árpád Dynasty fauna affected the right femur of a cat *(Felis catus* L., 1758). Signs of a grave but healed fracture can be seen on this bone, showing that the animal survived the trauma (Fig. 10.3). The fracture healed with slight dislocation and callus formation (Tasnádi-Kubacska 1960, 93), but as the other part of the bone tissue is healthy, it is likely that it was not an open fracture and could thus heal and without infection. Due to the dislocation, the animal must have had a limping right leg afterwards, despite the fact that the fracture had healed.

Medieval St. George Square, located in a quarter of the city close to the Royal Palace, also played an important military role during the sixteenth and seventeenth century Ottoman Turkish occupation. Although written records bear no mention of this area (Magyar 2003, 54), in addition to Turkish fortification works, the remains of smaller houses were also discovered here. The Islamic population of these quarters was largely of southern Slavic/Bosnian extraction (Fig. 10.1).

This paper (developed from a conference poster) considers the pathological animal remains from three important periods of the area: the late period of the Árpád Dynasty (second half of the thirteenth century), the Sigismund Period (turn of the fourteenth-fifteenth centuries) and the Ottoman Turkish Period (sixteenth-seventeenth centuries). Naturally, not all deformities can be presented within the framework of this study. Some of the bones show individual deformities, others are typical examples of anomalies observed on several specimens. However, this research is critical because it can provide more information about the environment of medieval people, the health in the medieval city and the nature of human-animal relationships.

## End of the period of the Árpád Dynasty (late thirteenth century)

An assemblage of 1,512 identifiable bones were recovered from a closed feature (pit 94 / 13) dating to the late thirteenth century (period of the Árpád Dynasty). This is important because, compared with other periods, relatively few archaeozoological

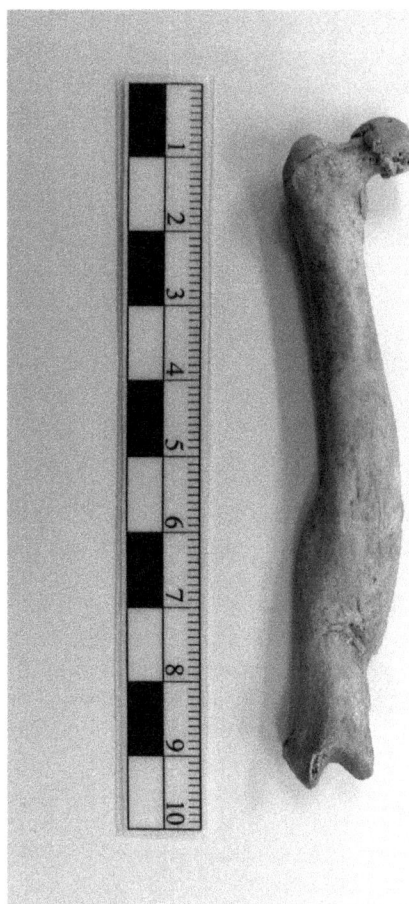

**Figure 10.3**: healed fracture on the femur of a cat (period of the Árpád Dynasty).

Figure 10.4: inflamed distal epiphysis of a cattle tibia (Sigismund Period).

## Sigismund Period (early fourteenth-fifteenth century)

From the Sigismund period, 22,748 bones were identified, of which six pathological specimens were recorded. First, an inflamed distal epiphysis of a cattle (*Bos taurus* L., 1758) tibia was identified (Fig. 10.4). The inflammation was probably caused by a fracture that affected the articulation. This trauma must have been neglected because of the lack of either attention or proper knowledge. As a consequence of this wound, however, the animal probably could not be used for working.

Another pathological bone from the Sigismund Period presented here is a fractured femur from a domestic chicken (*Gallus gallus* L., 1758; Fig. 10.5) that healed with a false joint (Tasnádi-Kubacska 1960, 97). This deformation must have had a definite effect on the animal's movement. However, since chicken require little space to move around it seems that this bird had a chance to survive for a long time, allowing the formation of a false joint. Unfortunately, other bones of the same skeleton were not discovered. Consequently, it cannot be established, how this anomaly affected the rest of the skeleton, especially the directly adjacent tibiotarsus.

Symptoms of rachitis (rickets) were observed on the right tibiotarsus of another domestic chicken (Fig. 10.6). Rachitis is a metabolic disease of the young caused by irregular bone mineralisation that results in retarded growth. This tibiotarsus not only shows bending, a typical symptom of this condition, but the distal end of the bone was also more robust (Horváth 1978, 364). The ossification of the epiphysis shows that the animal survived well after this condition developed. It was, however, probably hindered in its movement by the deformed limbs.

The right tibia of a dog originates from the same period. Symptoms of osteo-necrosis may be observed at the distal end of the bone presented in Fig. 10.7. It may be hypothesised that the animal's leg was lost due to some trauma. The concomitant inflammation was then isolated; thus, the rest of the bone was not affected. The necrotic parts were probably lost following the healing process. Several necrotic cavities may also be seen in this bone (Tasnádi-Kubacska 1960, 202). Amputation may be ruled out, because it would have left a callus of different morphology.

The left pelvis of a cat shows deformations caused by a poorly healed joint dislocation (Fig. 10.8). The *caput femoris* was forced out of the acetabulum and formed a new 'articular' surface of a reticular structure. While this new joint surface may have served its function, the cat's gait must have been heavily disturbed by this condition.

Bones of lagomorphs are rare among the quantities of animal bone found in the area. One of them, a right tibia, is of special interest owing to a fracture that healed with dislocation (Fig. 10.9). Fortunately, the matching left tibia from the same individual was also found. Since the epiphyseal plates of these two bones have not yet been ossified, one may conclude that the animal did not survive long after the healing of this trauma. Owing to the young age of this individual as well as on the basis of size and shape it is impossible to tell whether these bones originated from hare (*Lepus europaeus* Pall., 1778), caught easily because of this handicap, or domestic rabbit (*Oryctolagus domesticus* Erxleben, 1777), although the latter is known to have been introduced to Hungary only by the end of the Middle Ages (Bökönyi 1974, 335).

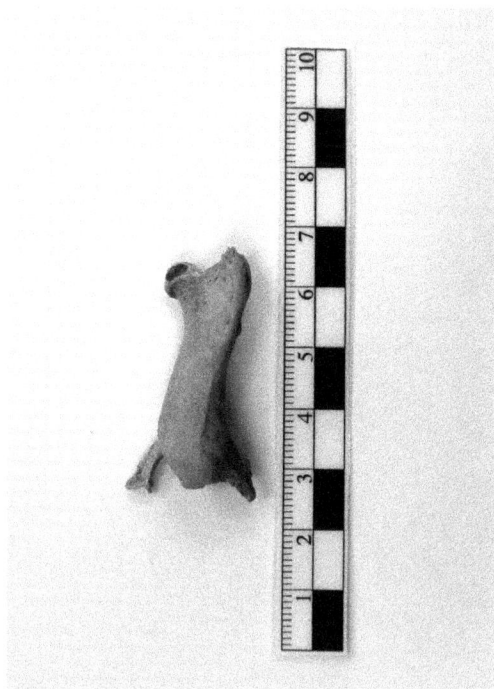

**Figure 10.5**: femur fracture in a domestic chicken that healed with a false joint (Sigismund Period).

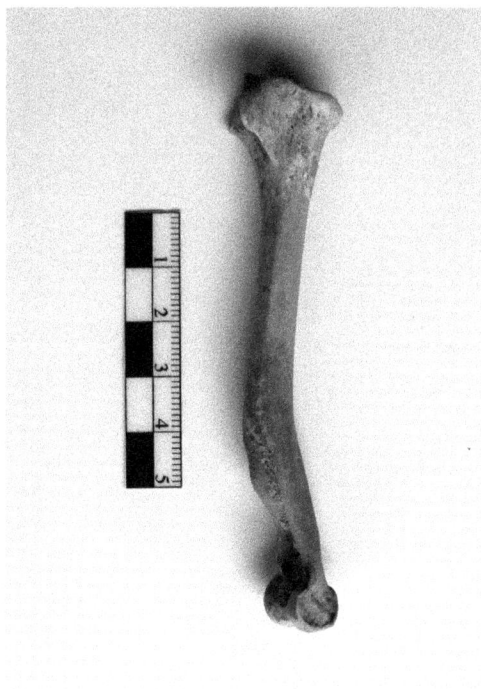

**Figure 10.6**: symptoms of rachitis on the right tibiotarsus of a domestic chicken (Sigismund Period).

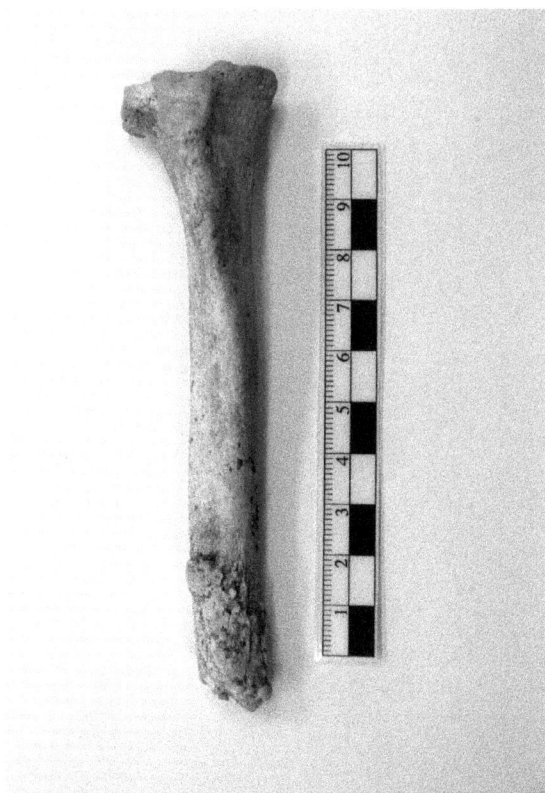

**Figure 10.7**: distal end of a dog tibia with symptoms of osteonecrosis (Sigismund Period).

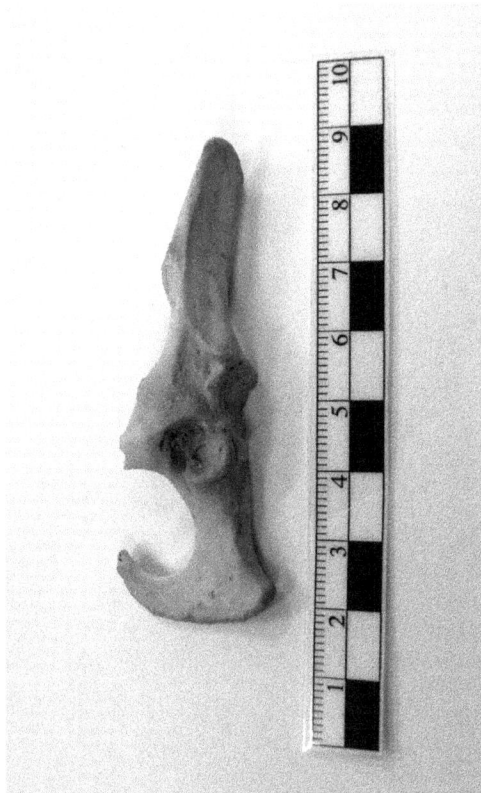

**Figure 10.8**: pelvis of a cat with deformations caused by a joint dislocation (Sigismund Period).

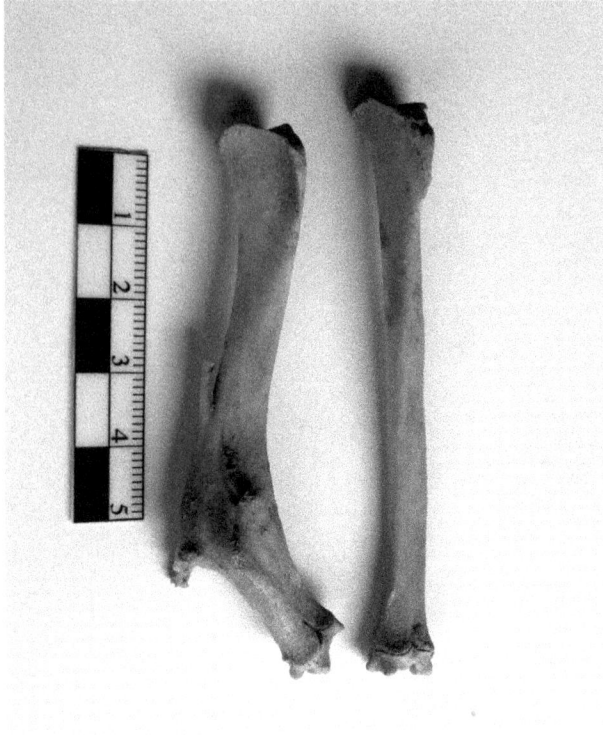

Figure 10.9: right lagomorph tibia with a fracture that healed with dislocation (Sigismund Period).

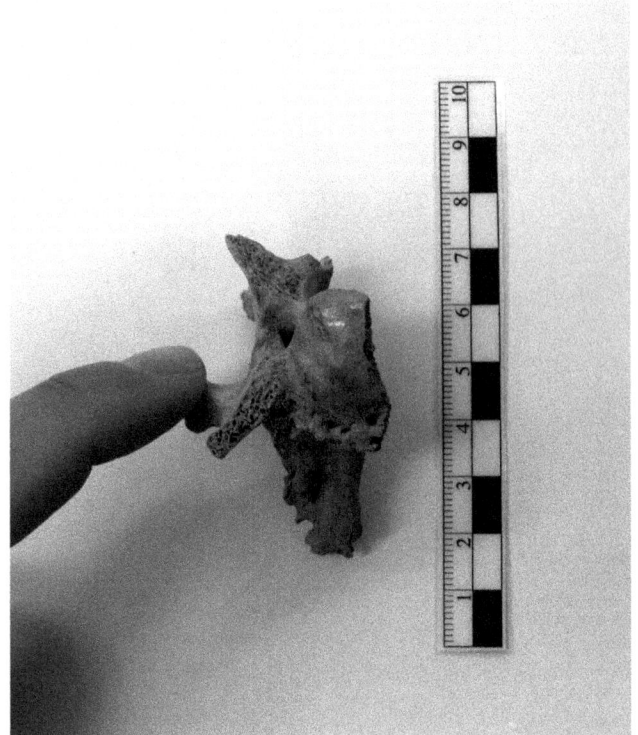

Figure 10.10: sheep cervical vertebra with symptoms of age-related osteoarthritis (Ottoman Turkish Period).

## Ottoman Turkish Period (sixteenth-seventeenth century)

Excavations brought to light 10,461 identifiable bones from the Ottoman Turkish period, but just two of these showed pathological symptoms. A sheep (*Ovis aries* L., 1758) cervical vertebra available from this period shows symptoms of age-related osteoarthritis (Fig. 10.10). This diagnosis is supported by eburnation on the cranial articular surface with irregular exostoses around its rim, as well as the broadening of the articular surface (Baker and Brothwell 1980, 115). Should the bone originate from a ram, repetitive strain injury (RSI) during intra-specific fighting in the mating season may have contributed to this condition.

Finally, the tibia of a young pig (*Sus scrofa* L., 1758) possibly damaged by some trauma (Fig. 10.11) was recovered. It is difficult to decide whether some metabolic condition resulted in the vulnerability of this bone or the damage was inflicted by simple mechanical force. It remains a fact, however, that the pig died, *i.e.* was slaughtered, before this disease was fully healed.

## Conclusions

An unusually great number, tens of thousands, of animal bones were brought to light during recent excavations in the Buda Castle. Symptoms of trauma and other diseases could only be observed on relatively few bones. In part, this may be due to the fact that seriously incapacitated domestic animals were slaughtered before

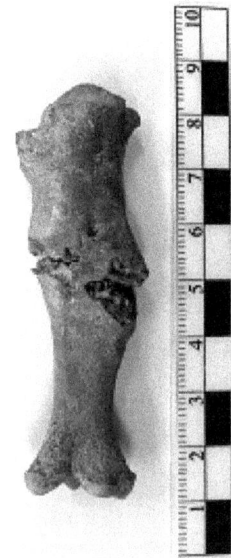

Figure 10.11: tibia of a young pig possibly damaged by trauma (Ottoman Turkish Period).

massive osteological symptoms of various diseases developed.

In spite of the great proportion of bones from adult/ mature cattle (Fig. 10.12), skeletal lesions attributable to draught exploitation were not observed. As the area the bones came from is situated close to the Royal Court, only nobility and wealthy members of society lived there. Such individuals and their families could probably afford to eat higher quality cuts of meat, not the stringy, tough meat from old draught cattle.

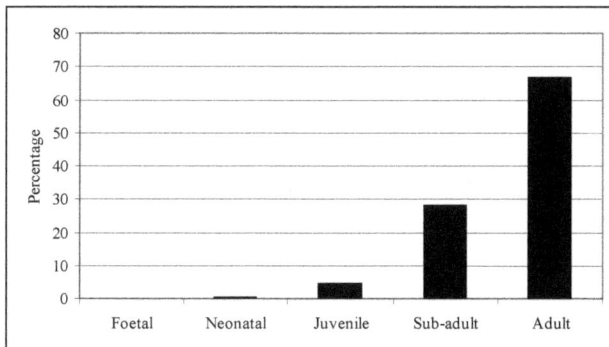

**Figure 10.12**: age distribution in cattle.

Regarding the remains of cats and dogs, pathological symptoms occur relatively more often, probably because these animals survived into old age and may have even received some treatment.

## Bibliography

Baker, J. R. and Brothwell, D. R. 1980. *Animal Diseases in Archaeology*. London: Academic Press.

Bökönyi, S. 1974. *History of Domesticated Mammals in Central and Eastern Europe*. Budapest: Akadémiai Kiadó.

Horváth, Z. 1978. *A háziállatok belgyógyászata*. Budapest: Mezőgazdasági Kiadó.

Magyar, K. 2003. A budavári Szt. György tér és környékének kiépülése. Történeti vázlat 1526-tól napjainkig. *Tanulmányok Budapest Múltjából* 31, 43-127.

Tasnádi Kubacska, A. 1960. *Paleopathologia. Az ősállatok patológiája*. Budapest: Medicina Kiadó.

Varga J., Tuboly, S. and Mészáros, J. 1999. *A háziállatok fertőző betegségei. Állatorvosi járványtan* II. Budapest: Mezőgazda Kiadó.

Zolnay, L. 1987. *Buda középkori zsidósága és zsinagógáik*. Budapest: Statisztikai Kiadó Vállalat.

## Authors' affiliation

Péter Csippán
Budapest History Museum
Szent György tér 2
1014 Budapest
Hungary

László Daróczi-Szabó
Budapest History Museum
Szent György tér 2
1014 Budapest
Hungary

# 11. Broken-winged: fossil and sub-fossil pathological bird bones from recent excavations

Erika Gál

## Abstract

*In this paper, a number of pathological lesions observed on avian remains are described. The skeletal parts reviewed derive from recently excavated archaeological sites representing various chronological periods in Hungary as well as a Late Pleistocene cave site in the Ukraine. Evidence of avian illnesses includes conditions of mechanical trauma as well as symptoms of infectious origin. Among the latter, advanced stage osteopetrosis was identified on the skeleton of a domestic hen (Gallus domesticus L., 1758.) from Budapest, found during the excavations of the medieval capital of Hungary. Early stage osteopetrosis was also recognised on the ulna of another hen from a seventeenth-century high-status site in northern Hungary.*

## Introduction

Birds are a special group of animals among vertebrates specialised to a different way of life. The majority of avian species are good flyers which, in addition to the evolution of plumage, have lead to the development of a modified skeleton, as well as respiratory and muscular systems. Birds generally have thin bones; the fragile pneumatic skeletal parts of many species serve to better utilise inhaled oxygen and decrease body density. However, such light bones are more susceptible to fracture and other injuries.

Bird remains usually only survive in small numbers both in palaeontological and archaeological deposits. One explanation for this is that the fragility of the avian skeleton and their small size leads to greater post-depositional loss than in the case of mammalian bones and teeth. However, many species of this class of vertebrates are difficult to hunt, both for predators and humans and the poor economic value and the strongly seasonal presence of many bird species motivated ancient people to concentrate on the hunting of large mammals. Nevertheless, the seasonal occurrences of molluscs, fish and birds have always attracted human interest. These animals offered not only a variety of additional animal protein but also a range of secondary products. Until the domestication of a few bird species and the spread of poultry keeping, fowling was the only source of eggs and feathers. Birds also had an important symbolic role in many societies; in medieval Europe, for example, hawking and the consumption of wild birds was an important marker of status (Albarella and Thomas 2002; Gál forthcoming).

Perhaps partly relating to their relative paucity, pathological specimens of bird bone have been rarely reported. Diseases observed on fossil bird remains were first reported in the late nineteenth century (Newton and Newton 1869). In Hungary, Kormos (1915) and Lambrecht (1916) pioneered the study of pathological animal bones. Tivadar Kormos (1915) mainly investigated the diseases of fossil cave mammals such as

the cave bear (*Ursus spelaeus* Rosenmüller, 1794) and cave hyena (*Hyaena spelaea* Goldfuss, 1823) excavated from the Late Pleistocene layers of caves located in the former territory of Hungary; however, the famous palaeo-ornithologist, Kálmán Lambrecht, focussed on bird bones and dedicated a chapter to the malformations of avian remains in his seminal hand-book (Lambrecht 1933, 880-889). The aim of this paper is to review the most characteristic pathologies visible on bird bones in recently excavated faunal assemblages.

## Material and context

The materials under discussion here derive from various sites. All avian remains are briefly summarised here to provide a context for the discussion of the pathological specimens. A summary of the origin of the finds and the nature and prevalence of the observed malformations is given in Tab. 11.1.

Two pathologically modified alpine chough (*Pyrrhocorax graculus* L., 1766) remains were found in the rich Late Pleistocene bone assemblage excavated from the Emine Bair Khozar "mega-trap" in the Crimean Peninsula, south-east Ukraine. The species list from this site also included two birds of prey, peregrine falcon (*Falco peregrinus* Tunstall, 1771), eagle owl (*Bubo bubo* L., 1758), and rock dove (*Columba livia* Gmelin, 1789), sky lark (*Alauda arvensis* L., 1758) and fieldfare (*Turdus pilaris* L., 1758). The special characteristic of this fossil bone assemblage is the dominance of alpine chough, and the very great number of skeletal parts from juvenile and sub-adult individuals (Gál 2005a). Out of the 696 alpine chough bones, however, only two showed pathological lesions. These two phenomena may be inter-related: there is less time for the development of chronic skeletal conditions in young animals, and the statistical probability of healed trauma is also smaller during a shorter life span.

| Species | Location | Date | Skeletal part | Pathology | Prevalence |
|---|---|---|---|---|---|
| Domestic duck (*Anas domesticus*) | Budapest – Teleki Palota (Hungary) | Middle Ages | Humerus (left) | Fracture healed with callus | 1/7 |
| Domestic hen (*Gallus domesticus*) | Budapest – Teleki Palota (Hungary) | Middle Ages | Sternum | Osteopetrosis | 1/179 |
| | | | Humeri | Osteopetrosis | 2/356 |
| | | | Radius (left) | Osteopetrosis | 1/62 |
| | | | Ulnae | Osteopetrosis | 2/253 |
| | | | Femur (left) | Osteopetrosis | 1/339 |
| | | | Tibiotarsi | Osteopetrosis | 2/402 |
| | Szendrő – Felsővár (Hungary) | seventeenth century | Ulna (left) | Early stage osteopetrosis | 1/56 |
| Crane (*Grus grus*) | Balatonkeresztúr – Réti-dűlő (Hungary) | Iron Age (La Tène) | Tarsometatarsus (right) | Ossified haematoma | 1/2 |
| Hooded crow (*Corvus* cf. *corone*) | Balatonlelle – Kenderföldek (Hungary) | Roman Period | Skull | Healed injury | 1/1 |
| | | | Ulna (left) | Fracture healed with callus | 1/1 |
| | Bajcsavár (Hungary) | Turkish Period | Skull | Healed injury | 1/1 |
| Alpine chough (*Pyrrhocorax graculus*) | Emine Bair Khozar "mega trap" (Ukraine) | Late Pleistocene | Ulna (left) | Early fracture (?) | 1/165 |
| | | | Tarsometatarsus (right) | Fracture healed with callus | 1/59 |

**Table 11.1:** summary of the studied avian bone pathologies.

The crane (*Grus grus* L., 1758) tarsometatarsus presented in this paper is part of a complete skeleton found in the La Tène feature B-1180 at the multi-period settlement Balatonkeresztúr – Réti-dűlő (Transdanubia, Western Hungary). A single tarsometatarsus (B-1295) from a juvenile crane as well as an incomplete skeleton (B-2073) of a chicken (*Gallus domesticus* L., 1758) and a single femur (B-1837) from an adult hen were also found in pits dated to the Iron Age (Gál 2007).

The Roman Period context, no. 1298, at the settlement Balatonlelle – Kenderföldek (Western Transdanubia, Western Hungary) contained an almost complete skeleton of a hooded crow (*Corvus* cf. *corone* L., 1758). Two bones of this specimen showed pathological conditions. Two remains from domestic goose (*Anser domesticus* L., 1758) and four bones from domestic hen were also found in coeval features at this site (Gál 2007).

The most spectacular pathological bird bones were excavated from a medieval context (C5/7) at the Teleki Palota (Palace) site in the Castle Area of modern-day Budapest, Hungary. The bone remains were most probably accumulated by the high-status residents of the site. Lanner (*Falco biarmicus* Temminck, 1825) must have been received as a gift or imported to Hungary since it did not form part of the local avifauna. Its presence, alongside other wild fowl such as partridge (*Perdix perdix* L., 1758), woodpigeon (*Columba palumbus* L., 1758) and magpie (*Pica pica* L., 1758) suggest a certain level of permanency in game hunting and fowling (Gál forthcoming). Poultry keeping, however, played an important role in animal husbandry as shown by the 789 goose and 604 chicken bones identified. The abundance of domestic fowl, representing 32% of the total number of animal bones indicates a strong interest in bird exploitation. Meat from domestic fowl as kosher food seems to have been especially important in the earlier, thirteenth-century period of the site, prior to the 1360 expulsion of the Jewish community from the area (Daróczi-Szabó 2002). In addition, the secondary products of these birds such as eggs, feathers and possibly the liver must have played an important role in the provision of the castle's inhabitants. Various pathologies were observed on some bones of a domestic duck (*Anas domestica* L., 1758) and on the incomplete skeleton of a chicken. Nevertheless, these ten bones represent only 1% of the total of domestic fowl remains recorded at this site.

A pathological hen ulna was found among the seventeenth-century bone deposits from Szendrő – Felsővár (Upper Castle) in North Hungary and may represent the same condition described in the hen skeleton above. According to written sources, this settlement was inhabited between 1599 and 1702. The bird bone assemblage also yielded the remains of domestic goose, hen, partridge, domestic pigeon (*Columba domestica* L., 1758), starling (*Sturnus vulgaris* L., 1758) and rook (*Corvus frugilegus* L., 1758). The tarsometatarsus of a large immature bird resembles turkey (*Meleagris gallopavo* L., 1758), which would provide one of the earliest occurrences in Hungary for this imported species. Although the number of bird bones was insignificant, the identified species indicate high-status consumption.

An injured skull of a hooded crow was found in the Ottoman Turkish Period fortress of Bajcsavár in Western Hungary. Various bird species were identified from the excavated bone assemblage that indicated a range of attitudes to this group of animals. Scattered remains of wild fowl such as white pelican (*Pelecanus onocrotalus* L., 1758), purple heron (*Ardea purpurea* L., 1766), coot (*Fulica atra* L., 1758) and wood-pigeon offer evidence for at least occasional hunting around the fortress. The relatively large number of bones from domestic species – goose, duck and hen – supports the idea that poultry keeping was an additional source of animal protein. Finally, remains of goshawk (*Accipiter gentiles* L., 1758), sparrow hawk (*Accipiter nisus* L., 1758), tawny eagle (*Aquila rapax* Temminck, 1828) and hooded crow suggest that a few bird species may have been kept as hobby animals, some of them possibly used in falconry (Gál 2005b).

**Figure 11.1**: healed fracture with callus formation on the tarsometatarsus of an alpine chough (Emine Bair Khozar "mega-trap").

## Analysis of pathological bones

### Fractures

Fractures are diverse and of varying degrees of severity; thus, healing may occur in many different ways. In the case of small fractures the two parts of the broken bone tend to be only slightly dislocated, therefore, the healing is rapid and does not cause spectacular changes in the shape of the skeletal parts. A spindle-shaped callus formation can be observed on the left tarsometatarsus of the Alpine chough excavated from the Emine Bair Khozar "mega trap" (Fig. 11.1). The fracture resulted in the shortening of the leg bone. Its greatest length is 35.1 mm, while measurements on 22 contemporary individuals gave the range of 41.8 – 48.9 mm with a mean value of 45.4 mm for this size (Tomek and Bocheński 2000, 65). Similar pathological changes were observed by Tasnádi Kubacska (1960, 165, Fig. 226) on the tarsometatarsus of a willow grouse (*Lagopus lagopus* L., 1758) from the Late Pleistocene Pilisszántó Rockshelter.

The malformation observed on the distal part of a broken ulna from the same Crimean bone accumulation may also be attributed to trauma occurring during the early stages of the bird's life (Fig. 11.2). Based on the undeveloped proximal epiphysis, this bone is attributed to an immature individual. It is hard to decide whether the distal part of the bone went missing ante- or post-mortem.

Figure 11.2: early trauma on the ulna of an alpine chough (Emine Bair Khozar "mega-trap").

The ulna of a hooded crow from the Roman Period settlement of Balatonlelle – Kenderföldek also exhibits a small fracture healed by callus that caused the shortening of the bone (Fig. 11.3).

A more severe form of fracture appears on the diaphysis of a left humerus from a domestic duck found at the aforementioned medieval site of Teleki Palota in Budapest. The two ends of the broken bone have been considerably dislocated and the soft tissue caught between them probably slowed down the process of healing. The slipped bones were connected by a callus-bridge. According to the trace of burning observed on the caudal aspect of the diaphysis, the bird might have been roasted (Fig. 11.4). This fracture also resulted in the shortening of the humerus that now measures 94.7 mm.

This value is slightly below the ranges of both females (109.9 – 113.0 mm, n=2) and males (106.1 – 109.1 mm, n=2; Woelfle 1967, 81). A similar rupture has been described on the femur of a willow grouse from the Late Pleistocene cave site, Puskaporos Rockshelter in Hungary (Tasnádi Kubacska 1960, 97, Fig. 142).

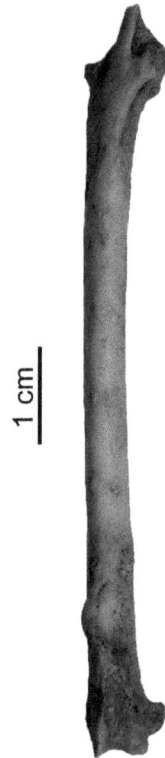

Figure 11.3: healed fracture on the ulna of a hooded crow (Balatonlelle – Kenderföldek).

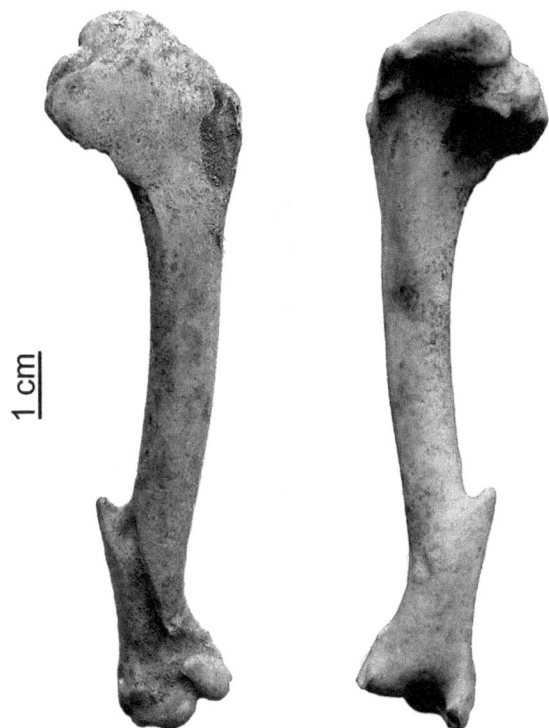

Figure 11.4: fractured domestic duck humerus exhibiting foreshortening and callus formation (Teleki Palota, Budapest). Left: cranial surface; right: caudal surface.

## Other traumatic injuries

Baker and Brothwell (1980, 82) have listed various causative agents for traumatic injuries. Some traumas observed on the bird bones under discussion here may be interpreted within these categories. Wounds on the skulls of hooded crows from the Roman Period settlement Balatonlelle – Kenderföldek and from the Turkish Period Bajcsavár (Fig. 11.5) may be related to intra-group conflict, which is well known among corvids. These, not very serious injuries, were most probably caused by the beaks of other birds and healed without complication.

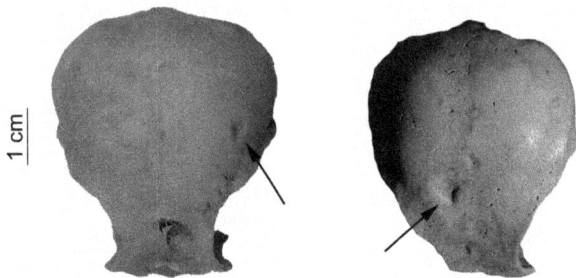

**Figure 11.5**: traces of traumatic injuries on the skulls of hooded crow (left: Balatonlelle – Kenderföldek; right: Bajcsavár).

Another type of wound was detected on the right tarso-metatarsus of a crane skeleton excavated from the Iron Age pit at the site Balatonkeresztúr – Réti-dűlő. The diaphysis shows a *c.* 6 cm long smooth bone swelling (Fig. 11.6) that may be an ossified haematoma. This kind of trauma-induced pathology is usually the result of a contusion that causes bleeding into a limited area of the sub-periosteal level of the bone. A similar example on a cervid metatarsus is given by Baker and Brothwell (1980, 83, Fig. 1).

**Figure 11.6**: ossified haematoma on a crane tarsometatarsus from an Iron Age feature at Balatonkeresztúr – Réti-dűlő.

## Osteopetrosis

Avian osteopetrosis is a typically identified disease of birds. In this condition, the action of the pathogenic agent results in the extensive thickening of the diaphysis of long bones both on the surface and inside the marrow cavity, while the epiphyses of bones remain more-or-less intact (Baker and Brothwell 1980, 61, Fig. 9). The most advanced archaeological case of this condition known by the author is from the medieval site of Teleki Palota in Budapest. The pathological bones collected included the sternum, the humeri, the ulnae, the left radius, the left femur and the tibiotarsi (Fig. 11.8). The most affected skeletal parts in this case were the sternum and the wing bones. Less affected by this deformation was the femur, while the tibiotarsi exhibited the smallest extent of the disease.

**Figure 11.7**: sternum, humeri, ulnae, radius, femur and tibiotarsi of domestic chicken (Teleki Palota, Budapest) with osteopetrosis.

Early stage osteopetrosis was observed on a chicken ulna (Fig. 11.8) found in the bone deposit of seventeenth-century date from Szendrő – Felsővár in North Hungary.

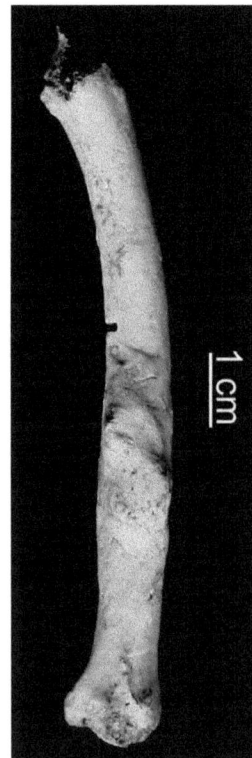

**Figure 11.8**: ulna of a domestic chicken (Szendrő – Felsővár) showing early-stage osteopetrosis.

## Discussion

Birds are predisposed in various ways to diseases that are reflected in the skeleton. The intense mobility of a great number of species probably creates one of the greatest risks. Birds often get into storms during peregrinations and migrations when they may crash against cliffs or the water surface, or collide with each other. Smaller wounds that heal in a few days usually allow the bird to join other individuals and continue the migration. Grave accidents, however, often lead to death.

Another occasion when injuries become frequent, particularly amongst male birds, is the mating season. Tasnádi Kubacska (1960, 159) describes two groups of birds that can be distinguished on the basis of fights during the reproduction period. Males in conflict suffer injuries both on the attacking parts, like the different elements of the wing, and on the body parts displayed for hits. Others, such as the cocks of ptarmigans (*Lagopus* sp.) and grouses (*Tetrao* sp.), which tend to simulate fights and use their beaks and legs, tend to be wounded on the chest and suffer fractures of the furcula (Tasnádi Kubacska 1960, Fig. 204).

There are groups of birds whose social attitude disposes the members to more frequent trauma than is the case of other species. Corvids are well known for their aggressive and quarrelsome behaviour which, together with their scavenging food habits, might partly explain human antipathy towards these otherwise intelligent birds (Bartosiewicz 2004).

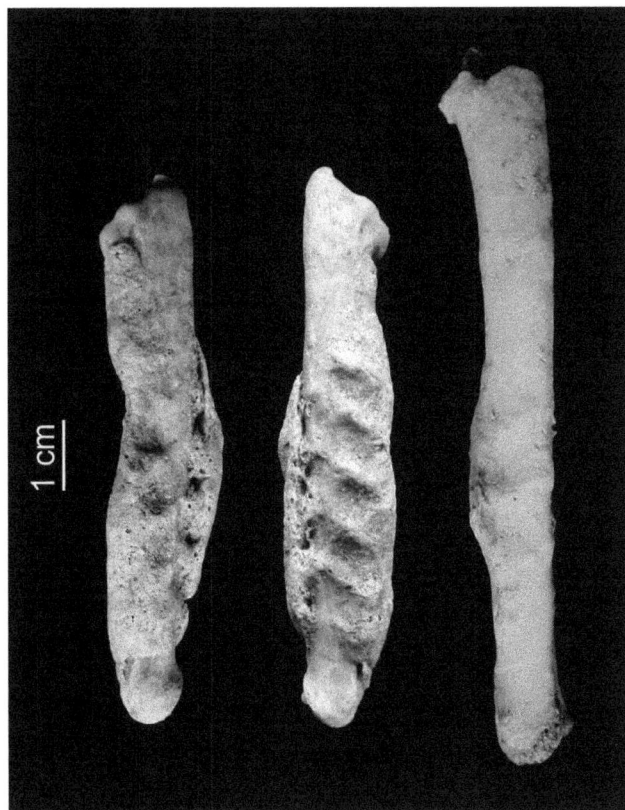

Figure 11.9: parallel outgrowths on the caudal surface of ulnae in domestic chicken (left: the pair of ulna from Teleki Palota, Budapest; right: ulna from Szendrő – Felsővár).

In addition to mechanical trauma, diseases caused by viruses and bacteria can also lead to skeletal deformation. The best-known condition is probably the chronic response initiated by osteopetrosis, which is caused by the avian leucosis virus (ALV) group, belonging to the virus family Retroviridae. Osteopetrosis results in extreme osteoblast proliferation and in the massive thickening of bones, especially those of the extremities. Although this condition was described by Baker and

Brothwell (1980, 61) as starting in the tibiotarsi of young chickens and then expanding with age to affect other skeletal parts, in the case of the bones available for study, these skeletal parts were less affected (Fig. 11.9). The sternum, the humeri, ulnae and the femur show the most serious deformations among the medieval chicken remains from the Teleki Palota excavated in Budapest (Fig. 11.7). This would suggest that the chest and proximal limbs were affected first, followed by the spread of the disease towards the periphery of the skeleton.

The similarity between the caudal surfaces of the ulnae on the specimens from Budapest and Szendrő suggests that the formation of parallel outgrowths, in addition to bone thickening. is indicative of early stage osteopetrosis in this part of the skeleton. The symmetrical development of these structures is most probably related to the location of the *papillae remigiales caudales* (not emphasised in galliformes) since the secondary remiges (feathers) are connected to the ulna by these *papillae* in living fowl.

We must take into account, however, that the present material is the first evidence for osteopetrosis in a small archaeological sample from Hungary and does not, therefore, permit far-reaching conclusions. An attempt at drafting the historical geography of this disease, at Roman and post-Roman sites in Britain is presented by Brothwell (2002).

## Conclusions

Although avian remains are usually present in small numbers in palaeontological and archaeological assemblages, they may illustrate a relatively broad range of bone pathologies. The most frequent condition is probably fractures in various stages of healing, the result of accidents and intra- or inter-specific conflicts such as fighting during the mating period or prey-predator relationship. Infections with diverse pathogenic agents largely affect poultry and spread both from bird to bird in the population and from hen to chick through the egg. Therefore avian domestication and husbandry, in addition to its evident economic advantages, may have lead to the expansion of different diseases, some of which can be studied through the bone remains and thus traced diachronically.

## Acknowledgements

This paper is the written version of the lecture presented at the ICAZ Animal Palaeopathology Working Group conference at Nitra (Slovakia) where the author was facilitated to attend by a Bilateral Fellowship between the Hungarian and Slovakian Academies of Science. I am grateful to the archaeologists and palaeontologists Szilvia Fábián, Gyöngyi Kovács, Dorottya B. Nyékhelyi, Gábor Serlegi, Gábor Tomka and Mátyás Vremir for inviting me to work on their excavated bird bone assemblages. Archaeozoologists László Daróczi-Szabó and Márta Daróczi-Szabó are thanked for selecting the avian remains from the vertebrate assemblages. The English text was revised by László Bartosiewicz and Richard Thomas. Thanks are due to the observations of an anonymous reviewer. This research was supported by OTKA Grant No. F 048818, and a Bolyai János Research Fellowship.

# Bibliography

Albarella, U. and Thomas, R. 2002. They dined on crane: bird consumption, wild fowling and status in medieval England, pp. 23-38, in Bocheński, Z. M., Bocheński, Z. and Stewart, J. R. (eds), Proceedings of the 4th Meeting of the ICAZ Bird Working Group, Kraków, Poland, 11-15 September, 2001. *Acta Zoologica Cracoviensia* 45 (special issue).

Baker, J. and Brothwell, D. 1980. *Animal Diseases in Archaeology*. London: Academic Press.

Bartosiewicz, L. 2004. Data on the culture history of crows (Corvidae) in the Hungarian Middle Ages, pp. 37-45, in Kovács, Gy. (ed.), *„Quasi liber et pictura". Tanulmányok Kubinyi András Hetvenedik Születésnapjára*. Budapest: ELTE Régészet-tudományi Intézet.

Brothwell, D. 2002. Ancient avian osteopetrosis: the current state of knowledge, pp. 315-318, in Bocheński, Z. M., Bocheński, Z. and Stewart, J. R. (eds), Proceedings of the 4th Meeting of the ICAZ Bird Working Group, Kraków, Poland, 11-15 September, 2001. *Acta Zoologica Cracoviensia* 45 (special issue).

Daróczi-Szabó, L. 2002. Animal bones as indicators of kosher food refuse from 14th c. AD Buda, Hungary, pp. 252-61, in O'Day, S. J., Van Neer, W. and Ervynck, A. (eds), *Behaviour Behind Bones: the Zooarchaeology of Ritual, Religion, Status and Identity*. Oxford: Oxbow Books.

Gál, E. 2005a. Taphonomic studies on Late Pleistocene avian remains from the Emine-Bair-Khosar „mega-trap" in South-East Ukraine, pp. 41-42, in Martinell, J., Domènech, R. and de Gibert, J. M. (eds), *Abstract Volume of the 2nd International Meeting TAPHOS'05, Barcelona (Spain), 16-18 June 2005*. Barcelona: Universitat de Barcelona - Fundació La Caixa.

Gál, E. 2005b. Vogelfunde aus der Festung von Bajcsavár, pp. 120-4 and 128, in Kramer, D. (ed.), *Auf Sand gebaut Weitschawar Bajcsa-Vár eine steirische Festung in Ungarn*. Graz: Historischen Landeskomission für Steiermark.

Gál, E. 2007. Bird remains from archaeological sites around Lake Balaton, pp. 79-96, in Zatykó, Cs., Juhász, I. and Sümegi, P. (eds), *Environmental Archaeology in Transdanubia*. Varia Archaeologica Hungarica 20. Budapest: Archaeological Institute of the Hungarian Academy of Sciences.

Gál, E. (forthcoming). "Fine feathers make fine birds": the exploitation of wild birds in the Medieval Hungary in Mulville, J. and Powell, A. (eds), *A Walk on the Wild Side. Proceedings of the ICAZ General Meeting, Mexico City, 23-28 August 2006*. Oxford: Oxbow Books.

Kormos, T. 1915. Fossilis csontokon észlelhető kóros elváltozásokról (Über krankhafte Veränderungen an fossilen Knochen). *Állattani Közlemények* 14, 244-78.

Lambrecht, K. 1933. *Handbuch der Paläornithologie*. Berlin: Verlag Gebrüder Borntraeger.

Newton, A. and Newton, E. 1869. On the osteology of the Solitaire or Didine bird of the Island of Rodriguez (*Pezophaps solitaria* Gmel.). *Philosophical Transactions of the Royal Society of London* 159, 327-62.

Tasnádi Kubacska, A. 1960. *Az ősállatok pathologiája*. Budapest: Medicina Könyvkiadó.

Tomek, T. and Bocheński, Z. M. 2000. *The Comparative Osteology of European Corvids (Aves: Corvidae), With a Key to the Identification of their Skeletal Elements*. Kraków: Institute of Systematics and Evolution of Animals Polish Academy of Sciences.

Woelfle, E. 1967. *Vergleichend morphologische Untersuchungen an Einzelknochen des postcranialen Skelettes in Mitteleuropa vorkommender Enten, Halbgänse und Säger*. München: Inaugural-Dissertation zur Erlangung der tiermedizinischen Doktorwürde der Tierärztlichen Fakultät der Ludwig-Maximilians-Univerität München.

## Author's affiliation

Institute of Archaeology
Hungarian Academy of Sciences
49 Úri
1014 Budapest
Hungary

# 12. Osteoporosis in animal palaeopathology

Monika Martiniaková, Radoslav Omelka, Mária Vondráková,
Mária Bauerová & Marián Fabiš

## Abstract

*While bones affected by osteoporosis are rarely reported in animal palaeopathology, numerous studies have documented the presence of this disease in ancient human skeletal remains. In such cases osteoporosis is often recognised by a significant reduction of trabeculae in spongy bone and/or the decreased thickness of cortical bone. However, it is generally known that palaeopathologists should employ histological analysis to avoid the possibility that diagenetic change is confused with the symptoms of osteoporosis. In this contribution, a suspected osteoporotic sheep (Ovies aries L., 1758) femur of medieval date is investigated using macroscopic and microscopic techniques. Histologically, a significant reduction in both spongy and compact bone was observed in the osteoporotic specimen compared with non-osteoporotic sheep femora. While various measurements of the Haversian canals disposed higher values in the suspected osteoporotic femur, no resorption cavities were identified in the microscopic structure. This could be partially explained by the age of the individual and/or the increased movement of the animal compared with humans. Irrespective of the cause of this phenomenon, histological analysis has confirmed the presence of early stage osteoporosis in the examined bone.*

## Characterisation of osteoporosis

Osteoporosis is the second most common pathological condition affecting the human skeleton in contemporary society following joint disease. In general, osteoporosis represents a systemic skeletal disease characterised by a reduction in the amount of bone or bone mass, and a deterioration of the microstructure of bone tissue, caused by a significant period in which bone resorption exceeds bone formation (Manolagas and Jilka 1995), resulting in increased susceptibility to fracture. A more clinically useful definition, recently proposed by the World Health Organization (WHO 1994), is based on measurements of bone mineral density. In this definition, bone mass in a given individual is compared to young normal individuals with maximum bone mass. The diagnosis of osteoporosis is based on a deviation exceeding 2.5 standard deviations (t score < -2.5).

## Histological aspects of osteoporosis

Osteoporosis, as manifested in the skeleton, involves two factors. The first of these is reduction in the number and diameter of trabeculae accompanied by a thinning of cortical bone. The second factor is a change in osteon remodelling, in which the rate of refill of resorption spaces is diminished and the thickness of the osteon wall is reduced, leaving a larger central Haversian canal (Ortner and Putschar 1981). The vertebral bodies usually show osteoporosis first and most severely. The normal dense vertical and transverse trabeculae become reduced. In this process the transverse trabeculae are more affected by osteoporosis than the vertical ones. The trabeculae are not only diminished in number but also in size. In the late stages of osteoporosis the few remaining vertical trabeculae may become reinforced (sclerotic atrophy). Although the bone is normal in matrix structure and mineralisation, the diminished quantity creates mechanical instability. Since the physiological turnover in cortical bone is much slower and less marked than in cancellous bone, long bones, which predominantly consist of cortical bone, are affected later and to a lesser extent by osteoporosis (Vyhnánek 1999). The resorption of the cortex proceeds in two patterns: endosteal resorption, which leads to increased diameter of the medullary cavity; and intra-cortical resorption, resulting in an increase in the numbers of unfilled or only partly filled Haversian resorption spaces. The latter process results in increasing porosity and lamination of the cortex (Ortner and Putschar 1981).

## Osteoporosis in past human populations

Since osteoporosis is an increasing problem among the elderly today, the extent to which it affected earlier human populations and also animals has become a growing focus of interest among palaeopathologists. Although in osteoporosis there is characteristically earlier and greater loss of trabecular bone, loss of compact bone plays an important role in increasing bone fragility (Boyce and Bloenbaum 1993). This, together with the fact that 80% of the skeleton comprises cortical bone (Polig and Jee 1987), means that its study is important if we are to gain an adequate understanding of osteoporosis.

Age-related bone loss is a global phenomenon and has been demonstrated in geographic areas spanning the climatic regions from the Arctic to the tropics. Ethnic and racial differences in the degree of bone loss have been observed. Afro-Americans suffer less from the complications of osteoporosis than do North American Caucasians, but this has been attributed to larger bone mass at mid-life (Aufderheide and Rodriguez 1998). In the eleventh–sixteenth century group from Wharram Percy (England), age-related loss of cortical bone, measured at the second metacarpal (Mays 1996) and the femur mid-shaft (Mays *et al.* 1998) resembled the pattern seen in modern populations, as did loss of bone mineral density in the proximal femur (Mays *et al.* 1998) and radius (McEwan *et al.* 2004). In general, women lose about 35% of their peak cortical bone mass and about half of their trabecular bone over a lifetime while men lose about two-thirds as much. In animals, similar studies have not been carried out.

Numerous studies have also documented the presence of osteoporosis in ancient human remains. Ericksen's (1976) study of three Native American archaeological populations found osteoporosis predominantly among females, and osteoporosis was found to afflict ancient Nubian women at an earlier stage than in the modern population studied by Dewey *et al.* (1969). Conversely, Józsa (1997) mentions that osteoporosis was a rare alteration before the nineteenth century in Hungary. The consequences of osteoporosis (proximal femur fractures, radius fractures and vertebral fractures) were also uncommon in both mummies and in archaic skeletal material from that country. Agarwal and Grynpas (1996) have hypothesised that senile osteoporosis was rare, or perhaps did not exist, in prehistory. They base their argument on the lack of reporting of osteoporotic fractures. However, Pfeiffer (2000) notes that fractures associated with osteoporosis are difficult to quantify. Compression fractures of the vertebrae are highly variable in their manifestation, and may be caused by factors other than systemic bone loss. Fractures of the femoral neck may also occur soon before death and thus the healing that signals a break as ante-mortem may be absent. Because fractures are not inevitable outcomes of reduced bone mass and because they are difficult to quantify, hypotheses about prehistoric osteoporosis would best be tested not through fracture frequencies, but through bone density measurements. Aufderheide *et al.* (1994) found severe osteoporosis with multiple vertebral crush fractures, kyphosis and even scoliosis in 5 out of 59 (8%) spontaneously mummified human remains of the Chinchorro culture that lived on Chile's northern coast more than 4,000 years ago. Four were women (ages 35, 50, 60 and 60±5 years) and one was a man (aged 40±5 years). Jósza and Pap (1996) have made a radiological examination of 341 adult skeletons from the tenth-twelfth centuries AD. They found the frequency of vertebral and femoral osteoporosis to be 7% among men over 61 years and 16.6% among senile females. Luzsa *et al.* (1988) radiographically detected only one case of osteoporosis, in a 39-year-old female, among 15 Hungarian royal skeletons (the frequency: 6.67%).

## Osteoporosis in animals

Osteoporosis in the aged animal has been attributed in part to sub-clinical vitamin D deficiency inducing secondary hyperparathyroidism and, in part, to decreased storage of IGF-1 and transforming growth factor β (TGF-β) within the bone (Plesker and Zwerger 2002). These deficiencies result in a decreased coupling between bone resorption and bone formation (Boonen *et al.* 1995). Other reasons for the occurrence of osteoporosis might be: recent pregnancy (DeRousseau 1985); social subordination (Shively *et. al.* 1991); hereditability/immobilisation (Pritzker and Kessler 1998); and ovarian hormones (Lundon and Grynpas 1993). In archaeozoological material, Baker and Brothwell (1980), in their study of animal palaeopathology, discuss the impact of osteoporosis on animal skeletons in the past; however, they note that such cases are very rare. In a study by Horwitz and Smith (2000), cortical bone thickness in sheep and goats showed a marked decrease from the Chalcolithic to the Early Bronze Age, and remains low in later periods. The authors suppose that this is a consequence of a negative calcium balance resulting from the combined effects of poor nutrition and intensive milking. It indicates that only in the Early Bronze Age were herds of sheep and goat exploited on a large scale for their milk (Smith and Horwitz 1984).

## Diagenesis – a source of false diagnosis

Reliable diagnosis is the basis for reconstructing the aetiology and epidemiology of diseases in ancient human populations and ancient animals. However, it is well known that diagnoses of diseases in ancient bones are not easily established. As a rule, palaeopathologists can only examine the vestiges of ancient diseases in dry bones, and no soft tissues or cells, which play an important role in pathological investigations on recent materials, can be studied to establish a reliable diagnosis or for comparative purposes (Schultz 2001).

In addition, archaeological bone is known to be affected under the earth by various factors (*e.g.* roots of plants, fungi, algae, bacteria, arthropods and their larvae, worms, protozoa, and chemical agents such as water and crystals). This process, diagenesis, is characterised by the transformation and/or destruction of bones by physical and chemical agents produced by the factors described above. Diagenesis occurs in the soil after the process of decomposition (Haskell *et al.* 1997; Rodriguez 1997). All these post-mortem factors produce damage that can falsely be diagnosed by palaeopathologists as lesions caused *intra vitam* by diseases (pseudopathology; Schultz 2001). For instance, compact bone can be destroyed by characteristic tunnel-like canals caused by the post-mortem growth of fungi or algae (Hackett 1981). These tunnels range in diameter from 5 to 10 µm and under electron microscopy appear empty with well-defined calcified walls, implying that collagen and mineral are both resorbed in the bone (Jans *et al.* 2004). Due to intensive growth of fungi, the tunnels can flow together and produce relatively large destruction holes that can be macroscopically mistaken for *intra vitam* osteoporosis or vestiges of a metastasising tumour (Schultz 2001).

It is difficult to predict from the gross appearance of a bone whether the tissue will prove to be diagenetically altered at the microscopic level. Still, some generalisations can be made. Histological preservation is likely to be poor if the burial environment has been moist and/or if the bone is easy to cut, 'chalky', and yields no smell of grease. The presence of purple (or any other colour) fungus within the cortex is a bad sign. Desert conditions, shell middens, and proximity to copper (Cu) artefacts tend to yield well-preserved bone. Well-preserved bone is relatively easy to prepare for microscopy, but if diagenetic alteration has occurred there is very little that can be done to improve the histological view. Because diagenetically altered bone does not transmit light as well as normal bone, it appears to be 'too thick' (Pfeiffer 2000). If possible, research protocols should accommodate diagenesis by allowing the measurement of those structures that are visible rather than a complete census of all structures that once existed in a bone cross section. The sizes of cortical structures like osteons and Haversian canals can be quite variable, and they are not normally distributed (Pfeiffer 1998). Therefore, it is necessary to assess a large number of structures to get a true sense of their central tendency.

As a reliable diagnosis is the basis not only of the study of case reports but also of the aetiology and epidemiology of disease in ancient populations, palaeopathologists should employ histological analysis to avoid false diagnoses. The necessary basis for such research is the knowledge of the general histology, histogenesis, and growth as well as pathophysiology of bone (Schultz 2001).

## Cadmium – a risk factor of osteoporosis

It is generally known that osteoporosis is a multi-factorial disease with dimensions of genetics, endocrine function, exercise and nutrition. Of particular considerations are the status of calcium (Ca), vitamin D, fluoride (F), magnesium (Mg) and other trace elements. Several trace elements, particularly copper (Cu), manganese (Mn) and zinc (Zn), are essential in bone metabolism as co-factors for specific enzymes (Saltman and Strausse 1993). Environmental exposure to cadmium (Cd) has been suggested as an environmental risk factor for osteoporosis in many studies, but the critical level of the exposure, the risk of bone damage, and the gender-dependent differences in the vulnerability to the bone injury at low and moderate environmental exposure are still unknown. Bone tissue is characterised by its markedly lower ability to accumulate Cd, compared to soft tissues such as liver and kidney (Habeebu *et al.* 2000; Ogoshi *et al.* 1992). Intensive skeletal formation is the stage of the most active Cd accumulation and the metal retained during growth seems to contribute considerably to its accumulation over a lifetime (Hunder *et al.* 2001; Ogoshi *et al.* 1992). In bone, Cd may be incorporated into hydroxyapatite crystals (Blumenthal *et al.* 1995; Christoffersen *et al.* 1988) or bind to bone proteins, including metallothionein - MT (Oda *et al.* 2001). The metal incorporated into hydroxyapatite crystals affects their properties and contributes to weakening of the bone strength, whereas that bound to proteins, other than MT, may influence the biochemical metabolic processes occurring in the bone tissue. MT has been suggested to protect against Cd-induced bone injury (Habeebu *et al.* 2000). Gonzáles-Reimers *et al.* (2003) determined the levels of Cd in bone samples of 16 pre-Hispanic inhabitants of Gran Canaria, 24 pre-Hispanic domestic animals (sheep, goat and pigs) from this island, eight modern individuals, and 13 modern domestic animals. Their results indicate that modern individuals showed higher bone Cd values than prehistoric ones. Values of prehistoric individuals did not differ from those of the prehistoric animals, but were higher than modern animals. A significant correlation was observed between bone Pb and Cd in the study (r=0.61, P<0.001).

## Analysis of a suspected medieval osteoporotic sheep femur

## Materials and Methods

Among the archaeozoological remains excavated from the site of Dubovany, western Slovakia, a suspected osteoporotic sheep (*Ovis aries* L., 1758) femur, dated to the eighth-ninth centuries AD, was recognised. The identification of this condition was based on the macroscopic characteristics of the bone. Firstly, the compact bone of the femur wall is thinner (Fig. 12.1) compared with femora from other sites where no osteoporosis was diagnosed (*e.g.* Čalovo, a Roman settlement in south-western Slovakia; Fig. 12.2). The spongy bone of the Dubovany femur is also reduced compared with the trabecular bone of other femora. Following the macroscopic examination of the Dubovany sheep femur, histological analysis was conducted. Thin sections of the suspected osteoporotic specimen and femora displaying no evidence of this disease, were prepared according to the procedures set out by Martiniaková *et al.* (2006) and examined under a light microscope using a digital CCD camera. The quantitative histological characteristics were assessed using the computer software Scion Image (Scion Corporation, U.S.). The following variables were measured: area; perimeter; and the diameter of Haversian canals and secondary osteons. Measurements were taken on all mature osteons that were not in a resorption phase and could clearly be outlined using the software.

## Results

As documented in Figs. 12.3 and 12.4, the area, perimeter and diameter of Haversian canals all appeared

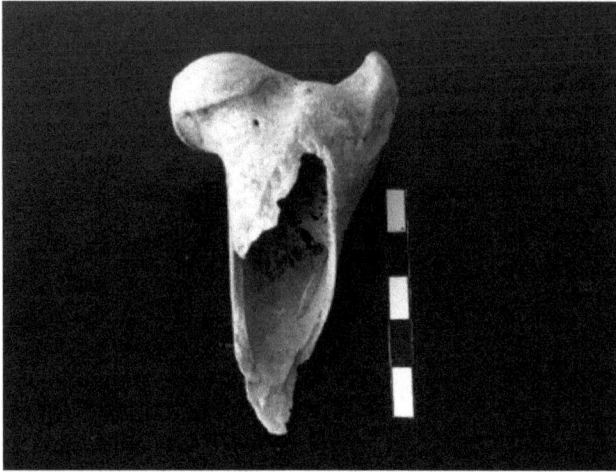

Figure 12.1: suspected osteoporotic sheep femur (scale bar = 5 cm).

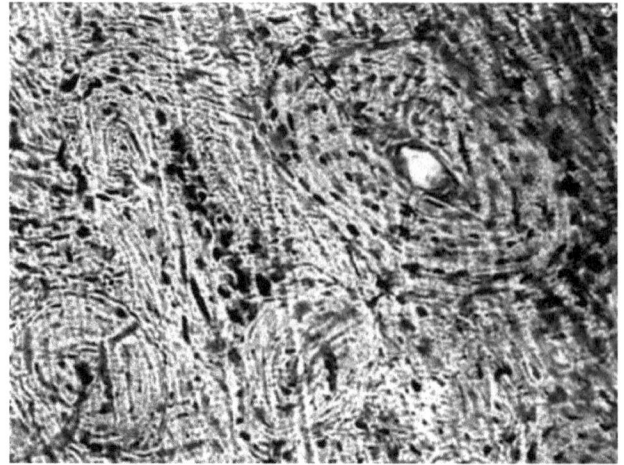

Figure 12.3: secondary osteons found in the microstructure of the suspected osteoporotic femur of sheep (magnification x 300).

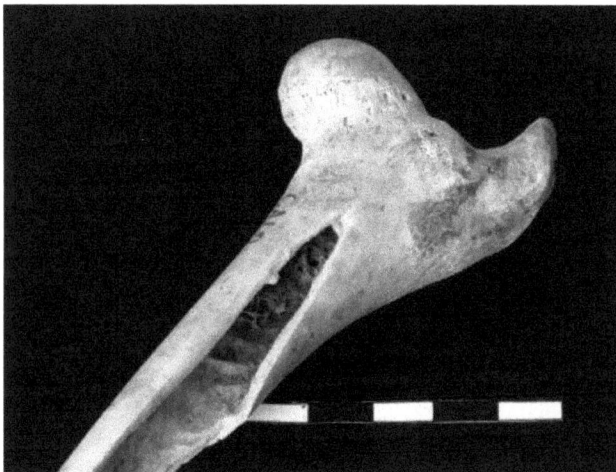

Figure 12.2: femur of a sheep without osteoporotic changes (scale bar = 5 cm).

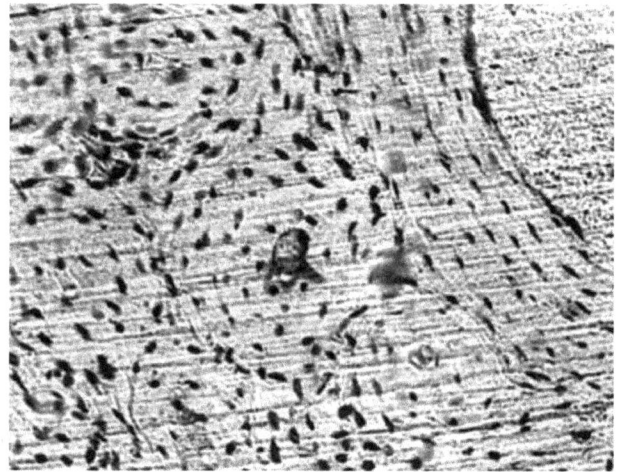

Figure 12.4: secondary osteons identified in the histological structure of a non-osteoporotic sheep femur (magnification x 300).

to be greater in the osteoporotic femur compared with the non-osteoporotic bones. Quantitative histological analysis the results of which are shown in Tab. 12.1, confirmed this fact. However, higher values of secondary osteons were identified in the non-osteoporotic samples. Similar results were found in a previous study in which the microscopic structure of human medieval osteoporotic and non-osteoporotic femora were analysed (Martiniaková 2006). However, no resorption cavities were observed in the microstructure of the suspected medieval osteoporotic sheep femur, either at the endosteal or the periosteal face. It is possible that this may have been caused by the age of the individual. According to macroscopic examination, the sheep was 12-15 months old; thus, osteoporotic changes, which result in many resorption cavities mainly at the endosteal border of the bone, may have little opportunity to develop.

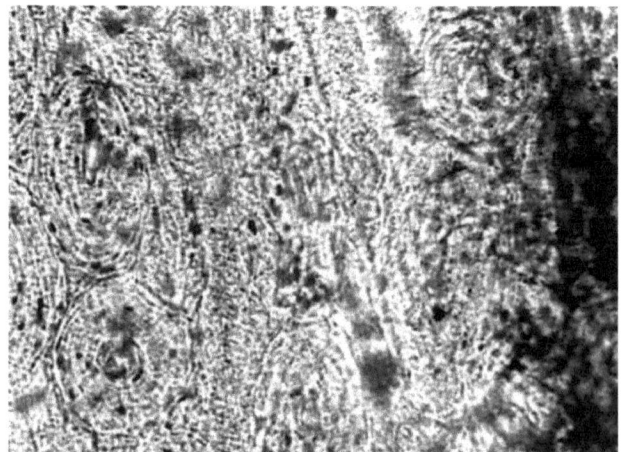

Figure 12.5: microbial attack on the microstructure of the suspected osteoporotic femur (magnification x 200).

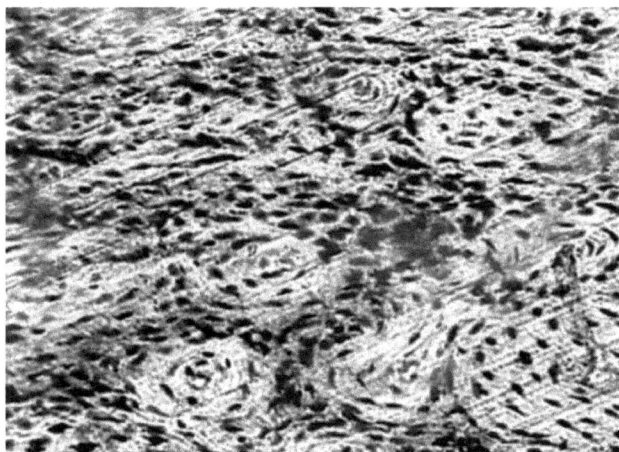

**Figure 12.6**: primary osteons identified in femoral microstructure of sheep (magnification x 200).

|  |  | Non-osteoporotic | Suspected osteoporotic |
|---|---|---|---|
| **Haversian canals** | Area | 617.73± 245.65 | 635.70± 246.59* |
|  | Perimeter | 68.95± 16.09 | 71.75± 29.99 |
|  | Diameter | 22.25± 8.78 | 27.18± 10.28* |
| **Secondary osteons** | Area | 20646.35± 8161.31 | 19848.14± 11957.69 |
|  | Perimeter | 421.73± 98.52 | 413.36± 97.75 |
|  | Diameter | 135.66± 32.91 | 128.45± 31.87 |

**Table 12.1**: results of quantitative histological analysis – all measurements are in microns. Key: * - $P<0.05$.

However, it could also relate to the increased movement of animals compared with humans, and consequently, the rate at which resorption spaces were refilled was not greatly diminished.

In the posterior parts of the periosteal surface of the suspected osteoporotic femur, evidence for bacterial destruction of bone micro-morphology was observed (Fig. 12.5). Other parts of the bone (anterior, medial, lateral) exhibited well-preserved bone microstructure. The extensive microbial damage identified near the periosteal surface of the posterior parts could be explained by the invasion of intestinal bacteria, which are known to play a role in the early stages of decomposition and perhaps even in bone alteration.

In addition to secondary osteons identified in the compact bone microstructure, many primary osteons were also found in the non-osteoporotic and osteoporotic sheep bone samples (Fig. 12.6). The results by Mori *et al.* (2005) and Martiniaková *et al.* (2007) indicate that thin sections from long bones of sheep have a mixed microstructure, with plexiform bone towards the surfaces of the specimens and Haversian bone mostly in the centre. Plexiform bone tissue consists, in general, of primary vascular canals (primary osteons) organised into

a regular, well-defined plexus. This research supports these observations.

## Conclusions

Osteoporosis is the most common type of bone disease. It is characterised by reduced bone mineral density, disrupted bone micro-architecture, and alterations in the amount and variety of non-collagenous proteins in bone. Osteoporotic bones are more at risk of fracture. In general, information about osteoporosis affecting the skeletons of non-human animals in the past has been rare. In this study, macroscopic and microscopic comparison of a medieval sheep femur with suspected osteoporosis with non-osteoporotic sheep femora was conducted. This revealed that the compact bone of the shaft of the osteoporotic femur was thinner compared with other non-osteoporotic bone samples. The spongy bone of that femur was also reduced compared with the spongy bone of other femora. Thus, the first factor of osteoporosis (reduction in number and diameter of trabeculae accompanied by thinning of cortical bone) was observed in the suspected osteoporotic bone. Histological analysis revealed that the measured variables of Haversian canals disposed higher values in the suspected osteoporotic medieval sheep femur. Higher values of secondary osteons were identified in the non-osteoporotic bone samples, although the differences were not confirmed statistically. In any case, larger Haversian canals were found in the microscopic structure of the osteoporotic specimen. While no resorption cavities were observed in the microscopic structure it is possible that this partly reflects the age of the individual and/or the increased movement of animals compared with humans. Thus, in terms of the second factor of osteoporosis (a change in osteon remodelling, in which the rate of refill of resorption spaces is diminished and the thickness of the osteon wall is reduced, leaving a larger central Haversian canal) is not apparent in this osteoporotic bone sample. Irrespectively, histological analysis confirmed an early stage of the disease in the examined bone.

In order to obtain more reliable results of compact bone tissue microscopic structure, more samples of ancient animals with and without osteoporosis require analysis. In so doing, we believe that all variables of the Haversian canals would exhibit higher values in osteoporotic animals; non-osteoporotic individuals should contain larger secondary osteons.

In general, archaeological bones are affected under the earth by various factors. Nevertheless, alterations caused *intra vitam* by disease or other living conditions can clearly be differentiated by light microscopy from changes due to post-mortem reactions. In this study, a microbial attack on the microstructure of the suspected osteoporotic femur was observed at the posterior parts of periosteal bone surface. Other parts of the femur disposed very well-preserved bone microstructure. It is generally known that histological condition can vary greatly within the same bone and within the different bones of the same individual. The results by Jackes *et al.* (2001) suggest that the foci of destruction are not randomly oriented with regard to bone structure, they are oriented mainly around the circumferential lamellae in cross section. The extensive microbial damage found near the periosteal surface of the suspected medieval osteoporotic femur of the sheep supports this assertion.

Environmental exposure to cadmium has been suggested as an environmental risk factor for osteoporosis in many studies. According to González-Reimers et al. (2003) modern individuals (including animals) showed higher bone Cd values than prehistoric ones. In general, cadmium is widely used in industry and is present in number of agricultural products. This could be a reason why osteoporosis is more frequently recognised in modern bone samples. We believe that a higher level of Cd will be determined in our suspected osteoporotic femur in comparison with sheep femora from other sites where no osteoporosis was diagnosed. This is the subject of current research. With further information regarding the association between Cd levels in bone and the manifestation of osteoporosis we will be able to identify the critical level of the exposure to Cd in the sheep with the suspected osteoporotic femur.

## Acknowledgements

The authors thank Prof. Dr. Bernd Herrmann (Institute of Zoology and Anthropology, Georg-August University, Göttingen, Germany) for allowing them to make the thin sections from the investigated femora in the laboratory of the Institute. This study has been supported by the grants KEGA 3/3181/05 and KEGA 3/4032/06 (Ministry of Education, Slovakia).

## Bibliography

Agarwal, S. C. and Grynpas, M. D. 1996. Bone quantity and quality in past populations. *Anatomical Record* 246, 423-432.

Aufderheide, A. C., Johnson, E. and Langsjoen, O. 1994. Health, demography, and archaeology of Mille Lacs Native American mortuary populations. *Plains Anthropologist* 39, 251-375.

Aufderheide, A. C. and Rodriguez, M. C. 1998. *The Cambridge Encyclopedia of Human Palaeopathology*. Cambridge: Cambridge University Press.

Blumenthal, N.C., Cosma, V., Skyler, D., LeGeros, J. and Walters, M. 1995. The effect of cadmium on the formation and properties of hydroxyapatite in vitro and its relation to cadmium toxicity in the skeletal system. *Calcified Tissue International* 56, 316-322.

Boonen, S., Aerssens, J., Broods, P., Pelemans, W. and Dequeker, J. 1995. Age-related bone loss and senile osteoporosis: evidence of both secondary hyper-parathyroidism and skeletal growth factor deficiency in the elderly. *Aging Clinical and Experimental Research* 7, 414-422.

Boyce, T. M. and Bloebaum, R. D. 1993. Cortical ageing differences and fracture implications for the human femoral neck. *Bone* 14, 769-778.

Baker, J. R. and Brothwell, D. 1980. *Animal Diseases in Archaeology*. London: Academic Press.

Christoffersen, J., Christoffersen, M. R., Larsen, R., Rostrup, E., Tingsgaard, P., Andersen, O. and Grandjean, P. 1988. Interaction of cadmium ions with calcium hydroxyapatite crystals: a possible mechanism contributing to the pathogenesis of cadmium induced bone disease. *Calcified Tissue International* 42, 331-339.

DeRousseau, C. J. 1985. Aging in the musculoskeletal system of rhesus monkeys: III. Bone loss. *American Journal of Physical Anthropology* 68, 157-167.

Dewey, J. R., George, J. A. and Murray, H. B. 1969. Femoral cortical involution in three Nubian archaeological populations. *Human Biology* 41, 13-28.

Ericksen, M. F. 1976. Cortical bone loss with age in three Native American populations. *American Journal of Physical Anthropology* 45, 443-452.

González-Reimers, E., Velasco-Vazquez, J., Arnay-de-la-Rosa, M., Alberto-Barroso, V., Galindo-Martin, L. and Santolaria-Fernandez, F. 2003. Bone cadmium and lead in prehistoric inhabitants and domestic animals from Grand Canaria. *The Science of the Total Environment* 301, 97-103.

Habeebu, S. S., Liu, J., Liu, Y. and Klaassen, C. D. 2000. Metallothionein-null mice are more susceptible than wild type mice to chronic CdCl2-induced bone injury. *Toxicological Sciences* 56, 211-219.

Hackett, C. J. 1981. Microscopical focal destruction (tunnels) in exhumed human bones. *Medical Science Law* 21, 234-265.

Haskell, N. H., Hall, R. D., Cervenka, V. J. and Clark, M. A. 1997. On the body: insects' life stage presence, their postmortem artifacts, pp. 415-448 in Haglund, W. D. and Sorg, M. H. (eds), *Forensic Taphonomy. The Post-Mortem Fate of Human Remains*. Boca Raton: CRC Press.

Horwitz, L. K. and Smith, P. 2000. The contribution of animal domestication to the spread of zoonoses: a case study from the southern Levant. *Ibex – Journal of Mountain Ecology* 5, 77-84.

Hunder, G., Javdani, J., Elsenhans, B. and Schumann, K. 2001. 109Cd accumulation in the calcified parts of rat bones. *Toxicology* 159, 1-10.

Jackes, M., Sherburne, R., Lubell, D., Barker, Ch. and Wayman, M. 2001. Destruction of microstructure in archaeological bone: a case study from Portugal. *International Journal of Osteoarchaeology* 11, 415-432.

Jans, M., Nielsen-Marsh, C. M., Smith, C. I., Collins, M. J. and Kars, H. 2004. Characterisation of microbial attack on archaeological bone. *Journal of Archaeological Science* 31, 87-95.

Józsa, L. 1997. The antiquity of osteoporosis. *Acta Biologica Szeged* 42, 75-80.

Józsa, L. and Pap, I. 1996. Az osteoporosis elöfordulása a X-XII századi magyarság körében. *Osteológiai Közl* 4, 126-129.

Lundon, T. and Grynpas, M. D. 1993. The long-term effect of ovariectomy on the quality and quantity of cortical bone in the young cynomolgus monkey: a comparison of density fractionation and histomorphometric techniques. *Bone* 14, 389-395.

Luzsa, G. Y., Gáspárdy, G. and Nemeskéri, J. 1988. Paleo-radiológiai tanulmány a székesfehérvari bazilika 15 csontmaradványáról. *Magyar Radiológia* 62, 39-50.

Manolagas, S. C., Jilka, R. L. 1995. Bone marrow, cytokines and bone remodeling. Emerging insights into the pathophysiology of osteoporosis. *New England Journal of Medicine* 332, 305-311.

Martiniaková, M. 2006. *Differences in Bone Microstructure of Mammalian Skeletons.* Nitra: Constantine the Philosopher University.

Martiniaková, M., Omelka, R., Grosskopf, B., Vondráková, M. and Bauerová, M. 2006. Prevalence of femoral osteoporosis in medieval humans from Dubovany cemetery (western Slovakia): macroscopic and microscopic examination. *Osteoporosis International* 17, (Suppl. 2), S55-S56.

Martiniaková, M., Grosskopf, B., Omelka, R., Vondráková, M. and Bauerová, M. 2007. Histological analysis of ovine compact bone tissue. *Journal of Veterinary Medical Science* 69, 409-411.

Mays, S. 1996. Age-dependent cortical bone loss in a mediaeval population. *International Journal of Osteoarchaeology* 6, 144-154.

Mays, S., Lees, B. and Stevenson, J. C. 1998. Age-dependent cortical bone loss in the *femur* in a mediaeval population. *International Journal of Osteoarchaeology* 8, 97-106.

McEwan, J. M., Mays, S. and Blake, G. M. 2004. Measurements of bone mineral density of the *radius* in a medieval population. *Calcified Tissue International* 74, 157-161.

Mori, R., Kodata, T., Soeta, S., Sato, J., Kakino, J., Hamato, S., Takaki, H. and Naito, Y. 2005. Preliminary study of histological comparison on the growth patterns of long bone cortex in young calf, pig, and sheep. *Journal of Veterinary Medical Science* 67, 1223-1229.

Oda, N., Sogawa, C.A., Sogawa, N., Onodera, K., Furuta, H., Yamamoto, T. 2001. Metallothionein expression and localization in rat bone tissue after cadmium injection. *Toxicology Lett*ers 123, 143-150.

Ogoshi, K., Nanzai, Y. and Moriyama, T. 1992. Decrease in bone strength of cadmium-treated young and old rats. *Archives of Toxicol*ogy 66, 315-320.

Ortner, D. and Putschar, W. 1981. *Identification of Pathological Conditions in Human Skeletal Remains.* Washington: Smithsonian Institute.

Pfeiffer, S. 1998. Variability in osteon size in recent human populations. *American Journal of Physical Anthropology* 106, 219-227.

Pfeiffer, S. 2000. Palaeohistology: health and disease, pp. 287-302, in Katzenberg, M. A., Saunders, S. R. (eds), *Biological Anthropology of the Human Skeleton.* New York: Wiley – Liss

Plesker, R. and Zwerger, C. 2002. Rickets, osteomalacia and osteoporosis in an indoor non-human primate facility. *Primate Report* 62, 69-78.

Polig, E. and Jee, W. 1987. Bone age and remodeling: a mathematical treatise. *Calcified Tissue International* 41, 130-136.

Pritzker, K. P. H. and Kessler, M. J. 1998. Diseases of the musculoskeletal system, pp. 415-460, in Bennett, B. T., Abee, C. R. and Henrickson, R. (eds), *Non-Human Primates in Biomedical Research: Diseases.* San Diego: Academic Press.

Rodriguez, W. C. 1997. Decomposition of buried and submerged bodies, pp. 384-389, in Haglund, W. D., Saltman, P. D. and Strause, L. G. (eds), The role of trace minerals in osteoporosis. *Journal of the American College of Nutrition* 12.

Schultz, M. 2001. Paleohistopathology of bone: a new approach to the study of ancient diseases. *Yearbook of Physical Anthropology* 44, 106-147.

Shively, C. A., Javo, M. J., Weaver, D. S. and Kaplan, J. R. 1991. Reduced vertebral bone mineral density in socially subordinate female cynomolgus macaques. *American Journal of Primatology* 24, 135.

Smith, P. and Horwitz, L. K. 1984. Radiographic evidence for changing patterns of animal exploitation in the southern Levant. *Journal of Archaeological Science* 11, 467-475.

Vyhnánek, L. 1999. Osteoporoza, p. 428, in Stloukal, M. (ed.), *Antropologie – příručka pro studium kostry.* Praha: Národní muzeum.

WHO 1994. *Assessment of Fracture Risk and its Application to Screening for Postmenopausal Osteoporosis.* WHO Technical Report Series No 843. Geneva: World Health Organisation.

## Authors´ affiliation

Monika Martiniaková
Department of Zoology and Anthropology
Constantine the Philosopher University
Nábrežie mládeže 91
949 74 Nitra
Slovakia

Radoslav Omelka
Department of Botany and Genetics
Constantine the Philosopher University
Nábrežie mládeže 91
949 74 Nitra
Slovakia

Mária Vondráková
Department of Zoology and Anthropology
Constantine the Philosopher University

Nábrežie mládeže 91
949 74 Nitra
Slovakia

Mária Bauerová
Department of Botany and Genetics,
Constantine the Philosopher University
Nábrežie mládeže 91
949 74 Nitra
Slovakia

Marián Fabiš
Necseyho 17
949 01 Nitra
Slovakia

# 13. Cranial perforations in Armenian cattle

## Nina Manaseryan

## Abstract

*This paper presents the analysis of 55 cattle skulls from five sites in Armenia. Thirty-four of these skulls exhibited cranial perforations affecting the parietal and occipital bones. Holes in the frontal bones of cattle are also presented for the first time. These holes varied widely in terms of their size, shape and location. The possible causes of these lesions are discussed.*

## Introduction

The study of pathological animal bones recovered from archaeological sites in Armenia is by no means an accomplished one. However, detailed studies of traumatic injuries and manifestations of bone disease in both sub-fossil and recent mammals have been shown to provide significant information regarding the nature of past human-animal interactions at different geographic and diachronic scales, from economic exploitation, to palaeo-environmental reconstruction and even prevailing social attitudes (*e.g.* Davies *et al.* 2005).

In order to understand the nature of past animal health and disease within Armenia, investigations of archived collections of animal bone that have been stored for many years are just beginning (Manaseryan *et al.* 1999). In this paper, the nature of those collections is discussed and further evidence for the presence of occipital perforations in cattle is presented.

## Materials

For over 70 years, scientists at the Institute of Zoology of the National Academy of Sciences of the Republic Armenia (NASRA) have travelled the globe to discover, describe and analyse the natural world and animal diversity. The Institute's osteological collections are based on material gathered during field studies of Armenian fauna and animal skeletons from zoos and related institutions. In the last 35-40 years the Institute has also conducted wide ranging research on vertebrate remains excavated from archaeological sites, which serve as a scientific resource for the study of the origin and evolution of domestic animals and the nature of past human societies. After species identifications are undertaken, the skulls and fragments of the post-cranial skeleton from archaeological sites are stored in collections specifically to facilitate future osteological studies. This material, which is unique in terms of its state of preservation, species diversity, and temporal coverage, has permitted the creation of a collection of sub-fossil animals that inhabited Armenia throughout its history. This collection contains 722 specimens of contemporary vertebrates (including 602 mammals and 120 birds) as well as over 10,000 samples of vertebrates from archaeological sites (skulls and post-cranial bones). Within this substantial collection of bones, there are many clear examples of animal pathology. While analysis of all pathological material within this depository is beyond the scope of the present paper, an observed anomaly present on a number of bovine skulls was considered worthy of further discussion since it is a condition that has been recently discussed in the literature (Baxter 2002; Brothwell *et al.* 1996; Manaseryan *et al.* 1999).

In total, 55 skulls were selected for analysis from the collection held at the Institute of Zoology. The majority of these derive from part of the Institute's collection gathered during expeditions undertaken by the Laboratory of the Vertebrates (1968-1972). They belong to natural bone burials from the Holocene era, concentrated on the south-western coast of Lake Sevan, near the village Akhkala (by the Ayrivan monastery; Fig. 13.1).

This site is situated along the wave-cut zone up to Noraduz cape and has a length of 500 m and a width of 50-60 m. Mezhlumyan (1972) indicated that the layers comprising the above-mentioned strata relate to palaeo-fluvial sedimentation of the Early Holocene. These layers contained the bones of animals lying within diagonal layers of sand. The species range of this 'faunal complex' is diverse and includes: red deer (*Cervus elaphus* L., 1758), wolf (*Canis lupus* L., 1758), fox (*Vulpes vulpes* L., 1758), wild boar (*Sus scrofa* L., 1758.), horncores of a sheep/goat (*Ovis aries* L., 1758 / *Capra hircus* L., 1758), and an abundance of the remains of a large Bos species, including European bison (*Bison bonasus* L. 1758; Manseryan *et al.* 1999).

In addition to this sample, domestic cattle (*Bos taurus* L., 1758) skulls from a number of archaeological sites were also studied. These derived from the Eneolithic settlement of Shengavit, Yerevan, in central Armenia, the second millinium BC site of Tsamakaberd just east of Sevan on the lake shore, an Early Iron Age cemetery at Lchashen, situated on the drained land 150-200 m from the Lake Sevan shore-line, and the Middle Bronze Age to Early Iron Age site of Loriberd, located three km from Stepanavan in the north-west of the country.

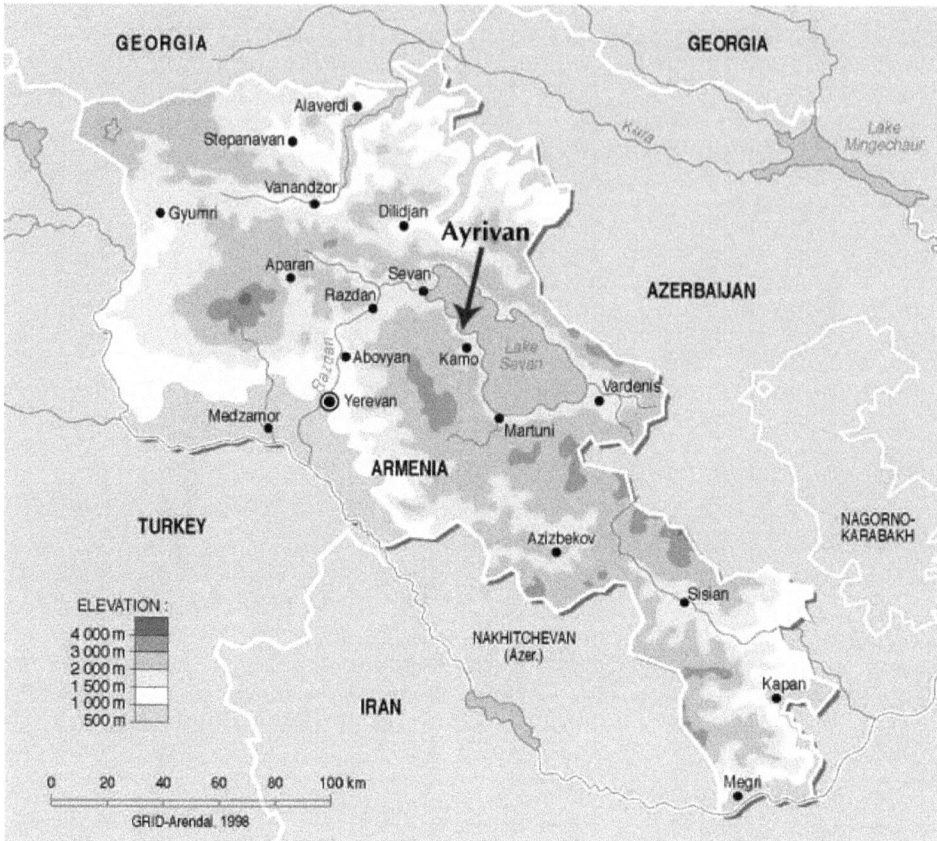

Figure 13.1: site locations.

## Analysis

Perforations of the frontal, parietal and occipital bones were identified and studied on 34 out of 55 skulls selected from the Institute's collection. Thirteen of the skulls derived from the 'faunal complex' from Akhkala; the remaining 21 were recovered from the archaeological sites noted above. The holes on nine skulls from the former complex and on four from the archaeological sites were located on the parietal bone: between the horns and occipital bones. A wide variety of permutations in form were observed. The perforations are sometimes single (e.g. Figs. 13.2-3) and sometimes multiple (Figs. 13.4-13.7). Moreover, some of the holes are clustered (e.g. Fig. 13.4-13.6), while others are further apart (Fig. 13.7-13.8).

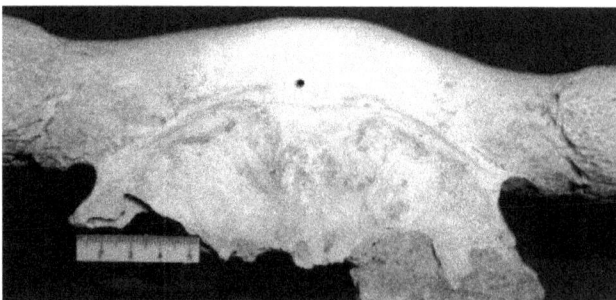

Figure 13.2: posterior view of a bovine skull fragment from Akhkala with one hole.

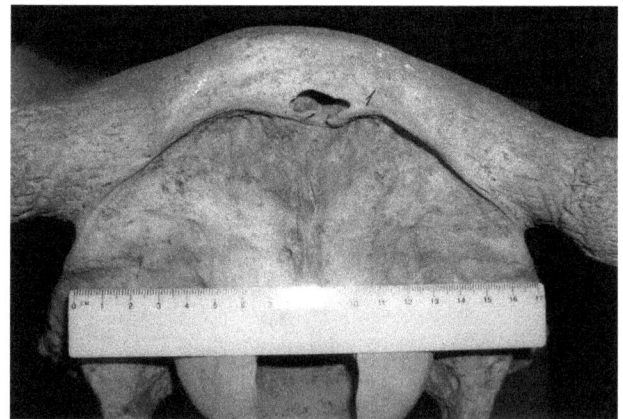

Figure 13.3: bovine skull from Lchashen with a single hole of irregular shape.

Some of the parietal perforations took the form of small circular holes 3-4 mm in diameter (Figs. 13.2 and 13.5), while others occur as oval or circular shapes of larger sizes, c. 5-10 mm in width (Fig. 13.5). The largest holes ranged from 13-19 mm in width (Figs. 13.3, 13.6 and 13.8). Perforations on the occipital bones ranged from 4-9 mm in width. Of particular interest is the presence of three circular areas of thinning above the occipital bone on a skull from Shengavit that have not yet resulted in perforation (Fig. 13.9).

Holes of another configuration occurred on the frontal bones of the facial part of the skull (Figs. 13.10-13.12). Like the parietal perforations, these occurred as both single and multiple holes of different shapes and sizes.

96

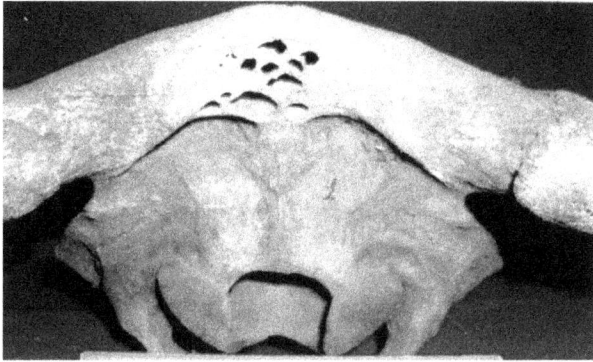

**Figure 13.4**: bovine skull from Lchashen with numerous holes of different sizes and shapes.

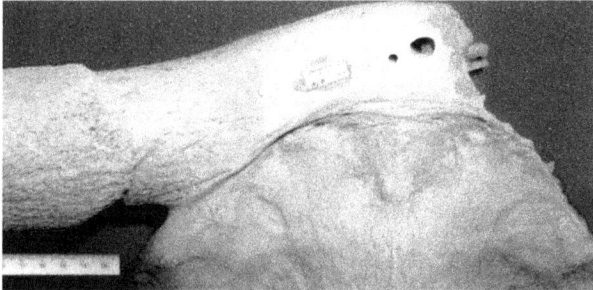

**Figure 13.5**: posterior view of a bovine skull from Akhkakla exhibiting two perforations.

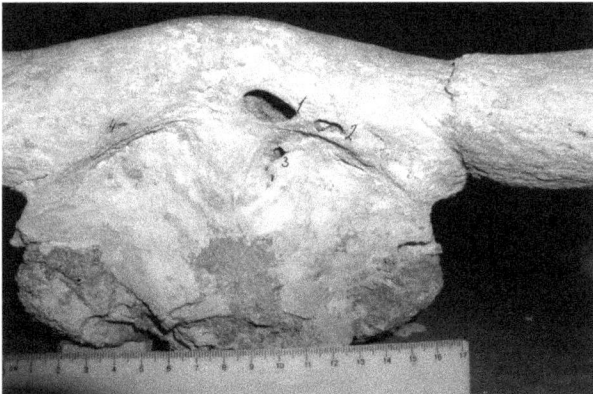

**Figure 13.6**: posterior view of a bovine skull from Akhkala with multiple perforations.

**Figure 13.7**: posterior view of a bovine skull from Loriberd exhibiting two perforations.

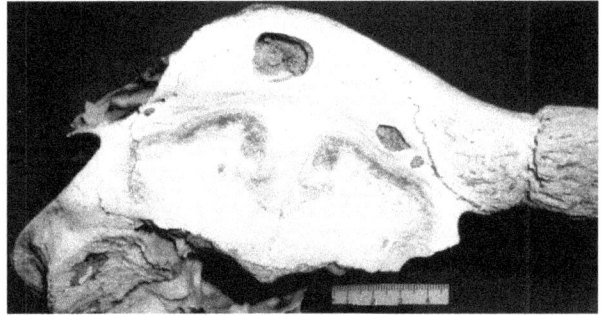

**Figure 13.8**: bovine skull exhibiting multiple holes of different shapes and sizes.

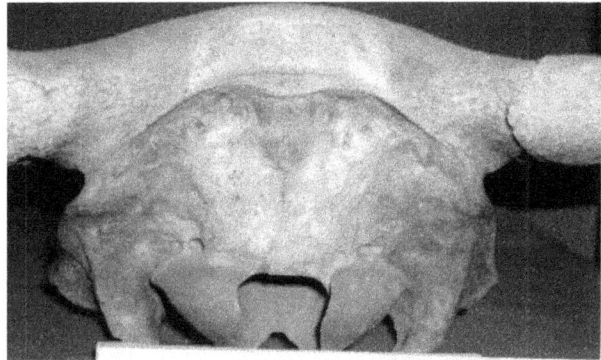

**Figure 13.9**: three circular areas of thinning over the occipital bone of a bovine skull from a Shengavit burial.

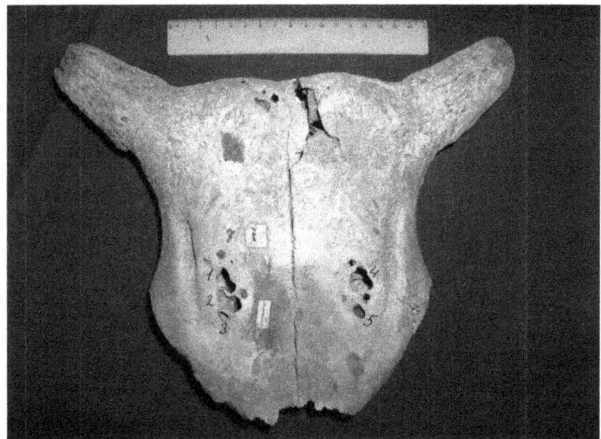

**Figure 13.10**: dorsal view of a fragment of a bovine skull from Tsamakaberd with many holes of different shapes and sizes.

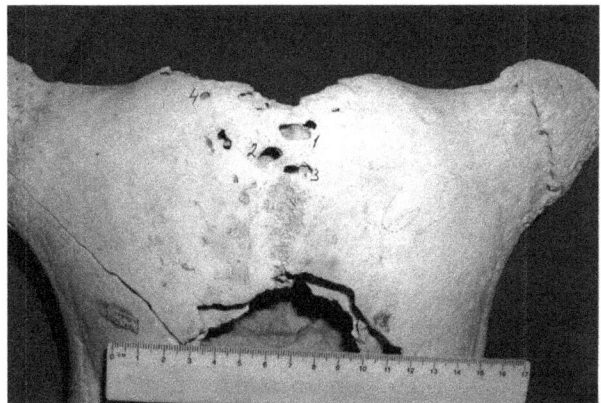

**Figure 13.11**: dorsal view of a fragment of a bovine skull from Akhkala with numerous holes of different shapes and sizes.

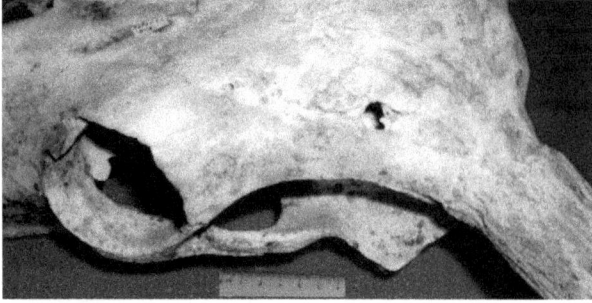

**Figure 13.12**: holes in the frontal bone of a bovine skull posterior to the eye socket.

They frequently occurred symmetrically in a variety of locations: adjacent to the frontal eminence; supra-orbital foramen; and posterior to the eye socket.

## Discussion

During life, animals may catch diseases, undergo periods of poor nutrition, be injured accidentally, during fights or even through hunting, all resulting in the development of skeletal pathologies. In Armenian archaeozoological literature these problems are seldom discussed; however, it is imperative that attempts should be made to explore the origin of these disorders.

Some of the perforations found in the occipital and parietal bones of the domestic cattle from the cemeteries at Lchashen and Loriberd burials could have resulted from the use of the animals for traction (see Brothwell *et al.* 1996). Unfortunately, however, it is too difficult to make any conclusions regarding the perforated crania from Akhkala. The fragmentary state of the skulls makes it impossible to determine whether they derive from domestic or wild bovids. That said, the presence of comparable perforations of the parietal and occipital bones in wild cattle elsewhere, notably aurochs (*Bos primigenius* Bojanus, 1827) and wisent, does argue against the traction hypothesis (Baxter 2002; Manaseryan *et al.* 1999) in favour of a congenital deformity (Brothwell *et al.* 1996).

This paper presents evidence for perforations of the frontal bones in cattle for the first time. Brothwell *et al.* (1996, 472) note, for example, that "we have not seen any cases of perforations outside the region of the external occipital protuberance or linea nuchae superior". As in the parietal and occipital regions, these holes were discovered singularly and as multiples of varying form.

The location of the burial sites from which these specimens derived, the shapes of the holes, the smoothness of the cavities observed argue against traumatic damage or taphonomic phenomena, such as penetration by tree roots, as causes. Some perforations on the frontal bone of the facial part of the skull could be obtained during the life as a result of trauma induced by humans or other animals. However, the similarity in form with the occipital and parietal lesions increases the likelihood that they too are the consequence of a congenital anomaly.

## Conclusions

Although the evidence is mounting, it is not possible at the present time to insist that the perforations of cattle crania are of congenital origin. The final solution, perhaps, would for anatomists, pathologists, veterinarians, anthropologists and archaeozoologists to collaboratively study these lesions. Only through joint efforts will we be able to expand out knowledge and evaluate the significance of pathological bones found from archaeological excavations.

## Bibliography

Baker, J. and Brothwell, D. 1980. *Animal Diseases in Archaeology*. London:Academic Press.

Baxter, I. L. 2002. Occipital perforations in a Late Neolithic probable aurochs (*Bos primigenius* Bojanus) cranium from Letchworth, Hertfordshire, UK. *International Journal of Osteoarchaeology* 12, 142-143.

Brothwell, D., Dobney, K. and Ervynck, A. 1996. On the causes of perforations in archaeological domestic cattle skulls. *International Journal of Osteoarchaeology* 6, 471-487.

Davies, J., Fabiš, M., Mainland, I., Richards, M. and Thomas, R. 2005. *Diet and Health in Past Animal Populations: Current Research and Future Directions*. Oxford: Oxbow Books.

Manaseryan, N., Dobney, K. and Ervynck A. 1999. On the causes of perforations in archaeological domestic cattle skulls (new evidence). *International Journal of Osteoarchaeology* 9, 74-75.

Mezhlumyan, S. 1972. *Palaeofauna from the Epochs of Eneolith, Bronze, and Iron Found on the Area of Armenia*. (in Russian). Yerevan: Publishing house of Academy Sciences of the Armenian SSR.

## Author's affiliation

Institute of Zoology
Armenian National Academy of Sciences
P. Sevak St. 7
0014 Yerevan
Armenia

www.ingramcontent.com/pod-product-compliance
Lightning Source LLC
Chambersburg PA
CBHW061009030426
42334CB00033B/3416